Algorithm Synthesis: A Comparative Study

D.M. Steier
A.P. Anderson

Algorithm Synthesis: A Comparative Study

With 11 Illustrations

Springer-Verlag
New York Berlin Heidelberg
London Paris Tokyo

D.M. Steier
A.P. Anderson
School of Computer Science
Carnegie Mellon University
Pittsburgh, PA 15213
U.S.A.

Library of Congress Cataloging-in-Publication Data
Steier, D. M.
Algorithm synthesis.
Includes bibliographical references.
1. Algorithms. I. Anderson, A. P. II. Title.
QA9.58.S74 1989 511'.8 89-19739
ISBN-13: 978-0-387-96960-2 e-ISBN-13: 978-1-4613-8877-7
DOI: 10.1007/978-1-4613-8877-7

Typeset by J.W. Arrowsmith Ltd., Bristol, England.

9 8 7 6 5 4 3 2 1

Preface

In early 1986, one of us (D.M.S.) was constructing an artificial intelligence system to design algorithms, and the other (A.P.A.) was getting started in program transformations research. We shared an office, and exchanged a few papers on the systematic development of algorithms from specifications. Gradually we realized that we were trying to solve some of the same problems. And so, despite radical differences between ourselves in research approaches, we set out together to see what we could learn from these papers. That's how this book started: a couple of graduate students trying to cope with The Literature.

At first, there was just a list of papers. One of us (D.M.S.) tried to cast the papers in a uniform framework by describing the problem spaces searched, an approach used in artificial intelligence for understanding many tasks. The generalized problem space descriptions, though useful, seemed to abstract too much, so we decided to compare papers by different authors dealing with the same algorithm. These comparisons proved crucial: for then we began to see similar key design choices for each algorithm.

The list of papers turned into a paper itself. But as we decided to do the comparisons in detail, we scaled back on the scope of the survey. We pruned our original list of over twenty algorithms down to the current seven, leaving out derivations for parsing and unification algorithms, among others. An original commitment to describe system implementations faded: there was too little published information available, especially on the system behavior for the examples we studied.

We presented some of our work at the ONR Workshop on Program Transformations in July of 1986, identifying key design steps for the quicksort algorithm, and giving authors' rationales for each step. The positive response to this presentation spurred us on, and we distributed a draft of the paper in the summer of 1987.

We made extensive revisions in response to the comments that came back. We abandoned some of our efforts to classify design activities in a uniform way. We did find it useful to classify the rationales for design decisions (into goal, selection and implementation knowledge), though we acknowledge that this classification is also imperfect. We had to reconcile ourselves to the fact that the field is still diverse enough that no survey could hope to satisfy all the requests for unambiguous definitions and complete summaries. At the very least, our efforts may convince others in the field to try to move towards consistent language for presenting their research.

The book would have been longer, but we decided to interrupt its writing to work on our own research. We would like to acknowledge the help of several colleagues in getting this far. Many people graciously agreed to wade through the preliminary drafts and gave us helpful suggestions on both form and content: Allen Newell, Bill Scherlis, Elaine Kant, Doug Smith, Michael Lowry, Guillermo Arango, David Gries, and Kathryn Swedlow. Our advisors, Allen Newell and Bill Scherlis, often discussed this work with us and helped us improve its structure (all the while tolerating the delay in our other projects). We are grateful to the sponsors of our fellowships, AT&T Bell Laboratories and the Schlumberger Foundation, and the School of Computer Science at Carnegie Mellon University, which provided space and support for the work. A final note of thanks goes to the authors of the papers we surveyed (and others in the field whose work we could not include), some of whom observed our progress and commented: "I'm glad somebody's doing this (and really glad it's not me)." This book is for them.

Pittsburgh
October, 1988

D.M.S.
A.P.A.

Contents

List of Figures

CHAPTER 1

Introduction

In the two decades since E.W. Dijkstra published a derivation of a synchronization algorithm [17], derivations for many algorithms have appeared in the computer science literature. The presentations are all motivated by a common belief that there is something to be learned from studying how algorithms might be designed from specifications. Yet, in looking at the derivations, one finds an almost overwhelming diversity: in choices of notations and examples, in levels of formality, even in basic underlying research assumptions. Does this diversity make the presentations completely incomparable, or is there a way to abstract from the individual presentations to obtain useful lessons for future research in this area?

As we will show in this monograph, the answer to this question is yes. Yes, because there are times when authors' goals are so radically different that the necessary information for meaningful comparison is simply not provided. But yes, also, because there are now several instances where a number of authors have published derivations for the same algorithms. In other subfields of computer science, standard "benchmark" problems have been crucial for facilitating communication ("dining philosophers" for synchronization mechanisms, and stacks and queues for data type specification are two instances). Now, by focusing on the algorithms for which multiple derivations exist, we have a similar opportunity for increasing mutual understanding in this area.

This book treats seven such algorithms as a foundation for a comparative study of the literature. It is not a comprehensive survey, and we have made no attempt to represent all the important schools of thought. For example, the school originated by Dijkstra [18], [28] is scarcely represented by the references to the work of Gries and of Gerhart in Chapter 6; the work of Constable and others [8], [14] on extracting programs from proofs is not represented at all. It is not our intention to write an introductory survey, and we warn readers unfamiliar with research in program synthesis or derivation against taking our sources as a definitive representation of the scope and value of this work.

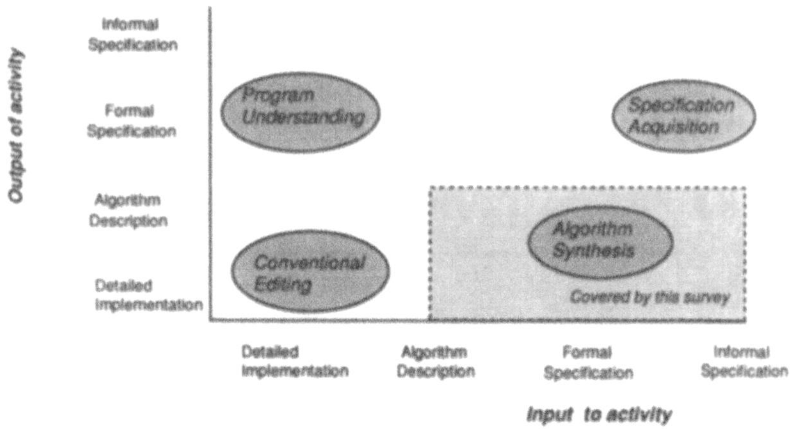

Figure 1.1. Inputs and outputs in software-related activities.

Our purpose is rather to discover if it is possible to establish some common ground for several rather disparate approaches found in the literature. The subfield of computer science covered by this literature bears a fairly clear relation to the rest of software research, as shown in Figure 1.1. We have plotted outputs of systems and methods of different software-related subfields as a function of their inputs, using level of abstraction from detailed implementation to informal specification as a rough indicator of position. For instance, some research treats the problem of formalizing an informal specification (e.g., the SAFE project [5]) or of abstracting from an implementation, perhaps for maintenance or documentation (e.g., the analysis component of the Programmer's Apprentice [60]). Our work is confined to the rectangular shaded region in the figure, the presentations that produce algorithms or implementations from problem statements that leave the complete computational method unspecified. The algorithms we studied are described informally below:

1. **Insertion sort:** Sort a given sequence by inserting elements into a sorted subsequence.
2. **Quicksort:** Sort a given sequence by selecting an arbitrary element, and placing all smaller elements to its left and all larger ones to its right.
3. **Cartesian set product:** Given two input sets, find all pairs such that the first element of the pair is in one set and the second element is in the other.
4. **Depth-first search in a directed graph:** Find all vertices reachable from a given vertex by traversing the graph depth-first, marking vertices already seen.
5. **Schorr–Waite graph marking:** Find all reachable vertices in a binary directed graph, a specialization of depth-first search to the Lisp garbage

collection problem. The maintenance of a parameter stack is avoided by manipulation of the graph representation.

6. **N-queens:** Place n (usually 8) queens on a chessboard subject to the constraints that they are all on different squares and none can capture any of the others.

7. **Convex hull:** Given a planar set of points, find the smallest convex polygon enclosing all the input points.

We chose the problems according to several criteria. Given the desire to find common examples, we limited our study to algorithms that have been treated at least twice in the literature. Because of this criterion, we excluded several unique derivations, including the one by Dijkstra mentioned above. We also excluded well-understood problems for which countless derivations exist, such as the problem of reversing a list. For the sake of generality, we wanted our sample to reflect a variety of problem domains. Our list now includes problems in sorting, set theory, graph theory, and computational geometry. Also, to avoid trivialization, we wanted the design of algorithms for these problems to cover a wide range of difficulty. Researchers have had varying degrees of success in formalizing and mechanizing the design of the algorithms we chose, though the majority could be taught in an undergraduate algorithms class.

Almost all previous surveys that have mentioned this literature (e.g., [20], [24], and [49]), have been organized around different systems or broad categories of techniques, rather than specific problems. In a few cases, specific derivations have been compared to other derivations as a means of discussing related work [48], but these comparisons have been focused on narrow domains. There has not been a broadly-based, systematic, comparison of research in the area organized around common examples. A prerequisite to such a comparison is a framework that is general enough to encompass presentations by authors with widely varying goals, methodologies, and notations. To motivate our particular framework, we will first distinguish between four styles of algorithm presentations:

1. **Rationalized descriptions:** Algorithms are traditionally presented mostly in their final form, with enough supporting theorems or argumentation shown to convince the reader of their correctness. Most algorithm design textbooks [2], [37] and most journal articles reporting the discovery of new algorithms follow this format.

2. **Design sequences:** A more detailed way of presenting an algorithm is to present the description as the product of a sequence of steps by which the algorithm could plausibly be developed, starting from a specification of the required behavior. The majority of such presentations describe a design as a linear sequence of steps, and do not provide detailed descriptions of alternative sequences that might result from applying other steps,

or the same steps in a different order. Design sequences have also been called "idealized design histories" [44]: idealized because they do not include the dead-ends one would expect a person to encounter in the design process. Most articles in program transformations [50], and at least one recent textbook on algorithms [19] display this presentation style.

3. **Design spaces:** Combining all possible design sequences that might result from a given specification defines a design space that in principle could be explored systematically. Approximations to this level of description are found in some of the reports describing derivations for families of algorithms [10], [26].

4. **Problem behavior graphs:** The final level of description supplements the description of a design space with a scenario or record of the process of exploring that space. Newell and Simon [47], adopted the term *problem behavior graph* (PBG) to describe such a record of the exploration process. This record includes not only which parts of the space were visited (including the extra effort not leading to a solution), but also the order in which the space was explored. In principle, a PBG could be used to describe the operation of an automatic design system, but one does not find this in the literature. Fragments of PBGs for human algorithm designers have been published [36].

As we have described them, the four methods of describing or summarizing a design fit into the view of design as search of a space, known in artificial intelligence as a *problem space* [46]. A problem space consists of a set of *states* and a set of *operators*. Each state is a structure representing a possible task situation. An operator produces a new state from an existing state. In the game of chess, a common example in AI, a problem space will have board positions as states and legal moves as operators. For a given problem space, a problem may be defined by specifying an initial state in the space and the desired final state or set of states to be reached by applying operators in the space. In chess, the initial position is the initial state, and the final state is a checkmate or drawn position. Problem space characterizations have also been used to build performance systems in several complex domains, so there is now a fairly well-developed technology base for building problem-solvers for tasks that have been analyzed this way.

Algorithm design is not nearly as well understood as chess, but we can approximate a similar analysis for it. As we wish to compare very different methods of presentation, we abstract away such features as specific programming languages or transformation rules, and speak, somewhat informally, of states and operators. We take designs at varying stages of refinement to be states. We take general principles or rules (which may be quite informal) that modify designs, for example, recursion removal, to be operators.

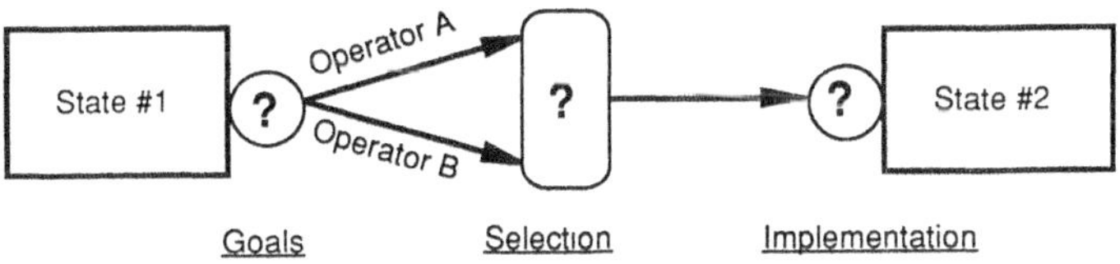

Figure 1.2. Different types of rationale used in operator proposal, selection, and implementation.

Together, these provide the basis for viewing the possible solutions to a design problem as a search space.

One cycle in this process of applying operators to states to produce new states is indicated in Figure 1.2. The question marks in the figure illustrate three other important elements that come into play when we consider the development of a particular solution, which we take to be elements of its *rationale.* First, in any task there must be a means for characterizing an acceptable solution. In designing an algorithm, one has criteria of good design in mind: efficiency, elegance, a particular target language or architecture, and so forth. Considerations like these are *goals*; they indicate what design objective is being served by applying an operator. Second, since design problems can only rarely be solved by blindly searching the space of possible solutions, one must often choose among alternative paths of development at a particular state. For instance, a designer might prefer a binary search to a sequential search because the worst-case number of comparisons is lower in a binary search. The relative time complexity of the two methods is *selection* knowledge; it makes one action preferable to another. Third, we describe operators in general terms, but the application of an operator to a particular design state often requires knowledge about the details of either the problem domain or the program itself. For example, case analysis is a general principle that requires knowing an exhaustive set of cases in particular. Such a set of cases is *implementation* knowledge; it justifies the details of the way a selected operator is implemented to produce a particular design.

When we consider the algorithm presentation styles discussed above in terms of problem spaces, they can be seen to form a sequence of increasing amounts of information about the search through the space. An algorithm description gives the point where the search terminates: the final state. A design sequence gives one path from an initial state (the specification) to a desired state (the algorithm) in terms of the intermediate states and the operators applied to move between states. The design space includes alternative paths. The problem behavior graph adds temporal information to give a record of the search process. The problem-space perspective provides an approach for understanding the presentations that is independent of the different authors' assumptions and systems. We wish to know what operators and states are in a design space, and for a particular derivation, what the knowledge is that leads to the design decisions defining a particular path through that space.

A full, detailed account of the design of any algorithm in these terms is not to be found in the literature, and such an undertaking would be formidable. Because the presentations we have studied leave much of the necessary information implicit, and due to the usual constraints on space (and reader patience), the analysis is less complete and exact than it might be.

Nevertheless, our feeling is that analysis at this level has provided us with useful information. We have asked two general kinds of questions: what specific knowledge is necessary for deriving each algorithm, and what is the nature of the algorithm design space? In examining the first question, we have focused on three issues. First, what knowledge about the problem is expressed in the specification, and in what form? Second, what derivational approaches can we distinguish as independent of the specific problem being solved? And third, what problem-specific elements of a derivation can we distinguish as capturing the "essence" of an algorithm? Our second general question is motivated by the hope that there is a benefit to be gained from representing algorithm design spaces explicitly in mechanized systems. By studying several derivations for the same problem, we discover part of the design space associated with the problem. This provides us with some case studies for research into questions about this design space, for example: what is its structure; how does the designer explore it; how can it be viewed as an explanation for a design; what are the patterns common to design spaces for different problems? These and other questions are important for us if we wish to explore the leverage that may be obtained by representing parts of the space in automated design support environments.

To help answer these questions, our analysis includes summaries for the individual derivations, observations about the collection of derivations for each algorithm, and general conclusions about the total collection of algorithms we studied.

The individual derivation summaries are grouped by the algorithm being designed. For each derivation, we start by giving the initial and final states, i.e., the problem specification and the algorithm, as described by the authors of the derivation. The heart of the individual derivation summaries is a chart showing the major states visited in finding a path from specification to algorithm; the operators used to get from state to state; and, where possible, a rationale for each operator application. In most cases, the authors of a derivation do not give a reason for every step in a development; rather than risk misrepresentation, we have chosen to limit ourselves to what is actually mentioned in the text of the original presentations. Each rationale is labeled as *goal*, *selection*, or *implementation* knowledge.

For each of the seven algorithms derived, there is a summary that has two components. The first component of these problem-specific comparisons is the listing of similarities and differences among specifications and results, with special attention being paid to the language used and the degree of commitment present in the data and control structures. The second component is a composite chart showing the part of the algorithm design space searched by the collection of derivations for that algorithm taken together. These composite charts show a generalization of the states visited and the operators applied (omitting the rationales). This format highlights points where the presentations converge. An interesting feature of these composites is how much overlap there is between presentations; even when the overlap is not present in the details, presentations often differ only in the commuting of several of the operator applications.

While we intend that the primary contribution of this study be in the individual analyses, we felt it useful to include observations on the collection of derivations as a whole. In the final chapter, we discuss issues raised by our survey in the areas of goals, languages, derivation structure, implementations, and presentation style.

CHAPTER 2

Insertion Sort

We begin by studying several derivations of algorithms to sort a sequence of data items into some specified order. Sorting is one of the most well-studied areas in algorithm design, both because one can produce a wide variety of useful algorithms, and because of the importance of sorting in practical applications: it has been estimated that more than 25 percent of all computer time is spent in sorting [37]. In this chapter we will study insertion sort, as an example of a very simple sorting algorithm.

Insertion sort is often the first sorting algorithm taught to students in programming. Instructors explaining insertion sort will usually begin by pointing out that it is the method of sorting that card players use. The idea behind the algorithm is to build up a sorted sequence by choosing an element and inserting it into the sorted sequence, extending the sorted sequence by one element each time. It is a simple algorithm to understand and implement, but is inefficient for large data sets since its performance is $O(n^2)$, where n is the number of elements to be sorted.

We analyzed the insertion sort presentations from Green and Barstow [26], Broy [10], Smith [54], and Clark and Darlington [13].[1] The presentations appeared in collections of several sorting algorithms: mergesort, selection sort, and quicksort, and sometimes others, were also derived. We will discuss quicksort derivations in Chapter 3. Though we have also compared mergesort and selection sort derivations, this monograph will not discuss them because the issues are directly analogous to those found in the insertion sort and quicksort derivations.

We begin the analysis with a comparison of the input specifications used as a starting point for each derivation. In the interest of brevity we have only included the top level for each of these specifications; that is, the definitions of predicates and terms used in the specifications were not included. The four specifications are:

---

[1] Clark and Darlington's derivation updates a more complex presentation using the same basic approach but a different syntax [15].

1. **Green and Barstow:** Though there is no formal specification, the synthesis begins with a goal to design a transfer sort, where the `transfer-elements` procedure is defined operationally, in terms of refinement rules in the PSI system.

2. **Broy:** The definition of the top-level sorting function specifies that an array is to be sorted:

   funct sort = (**array** A) **array**:
   that nat[1 ... upb(A)] **row m** X:
   hassameelements(A, X) $\wedge$ issorted(X);

   The declaration **nat**[1 ... upb(A)] **row m** X: specifies that the array X is to be considered as a row variable. The complete specification of sort also includes definitions for functions it calls, issorted and hassame-elements. In turn, hassameelements makes use of the function perm (permutation).

3. **Smith:** The specification requires that the input and output be lists of natural numbers:

   Sort: $x = z$ **such that** Bag: x = Bag: z $\wedge$ Ordered: z
   where Sort: List(N) → List(N)

   Implicit in this definition is an assumption of knowledge of what bags are and what it means for a list to be ordered.

4. **Clark and Darlington:** The specification is given as a collection of definitions and axioms about the domain, including a definition for the sort function:

   sort(x) $\Leftarrow y$ **such that** perm(x, y) $\wedge$ ord(y)

   The relations perm and ord are defined in terms of other primitives whose semantics are ultimately defined inductively on the structure of sequences.

The authors are in rough agreement on what constitutes an appropriate specification for the sorting problem. While they use very different notations, all four specifications contain (though not all explicitly at the top level) the same two notions that the output should be a permutation of the input, and that the elements of the output sequence should be ordered. Thus, the specifications all meet a primary requirement of specifications, that they be in terms appropriate to the application domain. However, an additional requirement is that specifications should not contain premature commitments to particular implementations. In these specifications, Smith commits to list representations of sequences in the specification, while Broy commits to arrays. Green and Barstow commit to use of `transfer-elements`, which is already an algorithmic design decision.

The four presentations of insertion sort compared here differ considerably in what they produce at the end of the synthesis. There are no established conventions for presentation of the code produced from a derivation, since authors stop when they have discussed the processes they feel to be "interesting." We attempt to present or summarize what the system or author produces when the problem is considered to be finished.

1. **Green and Barstow:** The synthesis described in [26] leads to an in-place sort with arrays for input and output, but the description of the algorithm is distributed throughout the presentation. Though code is not given in this article, the authors built a system that carried out the synthesis down to the level of code.

2. **Broy:** The last step in the derivation produces an iterative insertion sort expressed as two procedures in an ALGOL dialect of the wide-spectrum language CIP-L. Two versions are derived, one with a sequential insertion, and one with a binary insertion. The binary insertion version is presented below.

```
funct insertsort = (array A) array:
  ⌈(var nat [1 . . . upb(A)] row m X, var nat t) := (A, 2);
  while t ≤ upb(A) do (var nat min, var nat max) := (1, t − 1);
    while min ≤ max do nat j = (min + max)/2;
      if X[j] ≤ X[t] then min := j + 1
      □ X[j] ≥ X[t] then max := j − 1      fi od;
    (X, t) := (conc(insert(trim(X, 1, t − 1), min, X[t]),
                    trim(X, t + 1, upb(A))),
              t + 1)     od;
  X⌋
```

3. **Smith:** The derivation results in a recursive insertion sort expressed as three nondeterministic conditionals in a typed functional language (based on Backus' FP [3]). The first two conditionals express two divide-and-conquer algorithms, one for the top level sort function and one for inserting an element into an ordered sequence (the composition subprocedure). Isort takes a sequence of elements to be sorted as input. If the sequence is empty, it is returned directly as already sorted; otherwise, the sequence is split into its first element and the rest of the sequence. The first element is then inserted into the result of recursively sorting the remainder of the sequence. Insert takes an element and an ordered sequence as input. If the sequence is empty, the function returns a sequence containing only the element; otherwise, a conditional subprogram is called to decompose the input into smaller subproblems. The conditional compares the value of the element parameter to the value

of the first element of the sequence parameter. If we assume x_0 corresponds to the smaller element, x_1 to the larger element, and x_2 to the remainder of the sequence, the conditional returns a pair of the form $\langle x_0, \langle x_1, x_2\rangle\rangle$. The first parameter is then prepended to the result of recursively calling the insertion function on the second parameter (the nested pair).

Isort: $x \equiv$
  if $x = nil \rightarrow x$ $\|$
    $x \neq nil \rightarrow$ Insert $\circ$ (Id $\times$ Isort) $\circ$ [First, Rest]: x
  fi

Insert: $\langle a, x\rangle \equiv$
  if $x = nil \rightarrow$ Cons: $\langle a, x\rangle$ $\|$
    $x \neq nil \rightarrow$ Cons $\circ$ (Id $\times$ Insert) $\circ$ Decompose: $\langle a, x\rangle$
  fi

Decompose: $\langle a_0, x_0\rangle \equiv$
  if $a_0 \leq$ First: $x_0 \rightarrow \langle a_0, \langle$First: x_0, Rest: $x_0\rangle\rangle$ $\|$
    $a_0 \geq$ First: $x_0 \rightarrow \langle$First: $x_0, \langle a_0$, Rest: $x_0\rangle\rangle$
  fi

4. **Clark and Darlington:** The derivation yields two high-level expressions for a recursive insertion sort that works on lists. The second expression is a specification for insert, which is not elaborated.

   sort($[u]\langle\rangle x2$) $\Leftarrow$ insert(u, sort($x2$))

   insert(u, $y2$) $\Leftarrow z$ **such that** perm($[u]\langle\rangle y2$, z) $\wedge$ [ord($y2$) $\rightarrow$ ord(z)]

The difference both in language and degree of commitment in the resulting designs is readily apparent. The primary concern in the Clark and Darlington and Smith presentations is designing the algorithm (only the top-level sort in Clark and Darlington) while Green and Barstow and Broy go beyond deriving basic concepts of the insertion sort to produce reasonably efficient code in a conventional imperative programming language. This difference in emphasis shows up in the use of iteration or recursion, as well as the choice of arrays or lists to represent sequences. These considerations have substantial impact on the intermediate stages of the presentation as well, as will be shown in the discussion of the individual presentations.

2.1. Green and Barstow

Green and Barstow's insertion sort synthesis is part of a paper presenting a body of knowledge for use in synthesizing sorting programs. The knowledge was represented in a system that designed insertion sort automatically, but the emphasis in the presentation is on describing the knowledge used informally, rather than on giving the details of the representation or on rigorously proving the correctness of each step.

The presentation begins after it has been decided that a transfer sort will be designed. This kind of sort reorders the elements of a sequence by transferring them into a new sequence. More abstractly, the transfer paradigm consists of a process to produce elements from the input and another process to consume elements to produce the output. The top-level divide-and-conquer structure chosen for the sort reduces to a transfer program. An algorithm that produces one element at a time and inserts it into an ordered sequence is shown to fulfill the correctness conditions for an iterative program to satisfy the specification. At this point representation decisions are made and the algorithm is refined to be an in-place sort using arrays. The synthesis concludes with the design of an iterative insertion algorithm that finds the position in the sorted part of the array at which the element is to be inserted and shifts elements to make room for it.

State	Operator	Alternative	Rationale
Specification: transfer elements to ordered collection.			Given
	Choose top-level technique.		[Implementation] Assumed correctness conditions on split and join operation.
Divide-and-conquer program.		Generate-and-test permutations.	
	Choose relative sizes of subproblems produced by split.		
Singleton split.		Equal size split.	[Selection] Other path gives other algorithms.
	Hypothesize and test producer-consumer components.		[Goal] Want iterative program. [Implementation] Components satisfy consumer-producer correctness conditions.

(continued)

State	Operator	Alternative	Rationale
Iterative program with components producing and consuming one element at a time.			
	Decide where output stored.		[Goal] Space efficiency: $O(n)$ [Implementation] Initial and intermediate sets can be destroyed.
In-place sort.			
	Choose set representation.		[Implementation] If sets stored in contiguous regions of memory, can refer to set by array positions marking the boundaries.
Array representation of sets.		List representation of sets.	
	Refine shift technique		
Insertion of single element.		Selection.	[Selection] Minimize shifting; simplicity.
	Refine determination of position to insert.		
Linear search to find insertion position.		Binary search to find insertion position.	[Selection] Same number of moves in both; linear search is simpler.

Several features of this presentation merit some further comment. The first is the use of the transfer paradigm. Barstow has shown the transfer paradigm to be applicable to domains other than sorting, such as graph algorithms [6], and casting a system as a set of producers and consumers has clear uses in concurrent programming. The second is the substantial role of efficiency concerns in making design decisions. For example, the PSI coding system that automatically synthesized the insertion sort contains the knowledge that a transfer program for reordering data can be designed to put the burden of ordering the data either on the producer or on the consumer. An insertion sort has a simple producer and relies on the consumer to keep the output sorted. Later work on PSI in the LIBRA system made use of efficiency knowledge to make this decision [33]. When selecting the transfer order, LIBRA compared the cost of the partially refined insertion sort resulting from this step to a partially refined selection sort. Optimistic execution-cost estimates are slightly higher for selection sort, so the insertion sort path is chosen.

2.2. Broy

Broy's insertion sort derivation appears within a collection of derivations, whose purpose is to demonstrate that several well-known sorting algorithms can be derived systematically by the application of formal rules, each of which carries out a small part of the refinement. Though the report containing the sorting derivations does not describe an implementation, the goal of the CIP (Computer-aided Intuition-guided Programming) project, within which the research was conducted, is to build systems capable of supporting such derivations [9].

The states and operators in the table below show that the derivation follows the pattern of starting from a general characterization of the problem, and successively introducing commitments into the program by small steps to make the computation more efficient. The sorting problem is characterized inductively, leading to a recursive algorithm that sorts subarrays separately and then merges them. The major design decisions, to split off a single element at a time, and to insert this element into its proper position in the sorted subarray, are made in order to show how insertion sort can be derived as opposed to other sorting algorithms. A proof of correctness justifies the initial inductive sorting program, but proof does not play a role in the later stages of the derivation, as the transformations are correctness-preserving.

State	Operator	Alternative	Rationale
Specification: Output array is sorted, has same elements as input.		Specification: Output array is ordered, corresponding multiset equal to input multiset.	Given.
	Apply induction principle.	Refine to reduce number of inversions in array.	[Selection] Other paths give other algorithms. [Implementation] Correctness given by proof by induction on length of array.

(continued)

State	Operator	Alternative	Rationale
Algorithm sorts subarrays, then merges them.			
	Choose number of subarrays.		
Decomposition of array into two subarrays.		Decompose into more than two arrays.	[Selection] Other paths give other algorithms.
	Choose relative sizes of subproblems produced by split.		
Decomposition of array so one subarray has a single element.		Decomposition so both subarrays have nearly equal length.	[Selection] Other path gives other algorithms.
	Decide how to choose single element.		
Choose a single element at end of array.		Choose max or min element in array.	[Selection] Other path gives other algorithms.
	Decide how to use chosen element.		
Chosen element inserted into sorted subarray.		Subarray is partitioned around chosen element.	[Selection] Leads to insertion sort.
	Refine selection of position to insert chosen element.		
Insertion position found by binary search.		Insertion position found by linear scan.	[Selection] Reduce number of comparisons.
	Remove recursion.		[Goal] Efficiency.
Iterative program.			

The article assumes the guidance of the derivation by human intuition. Goals that would motivate the proposal and application of operators to a given state were left implicit. On the other hand, given that the operators often specialize some algorithmic structure and that other algorithms result from different specializations, there is more consideration of alternatives in this presentation than in other derivations.

2.3. Smith

Smith's derivation for insertion sort appears among a collection of derivations for sorting and other simple divide-and-conquer algorithms. The strategies presented for instantiating an abstract scheme for divide-and-conquer algorithms were implemented in the CYPRESS system. CYPRESS is being reimplemented and extended to cover other algorithm schemes.

The derivation begins by matching the sorting problem specification against the scheme for divide-and-conquer algorithms. The specification is decomposed into specifications of simpler subproblems, using a theorem giving the necessary conditions for the existence of a correct divide-and-conquer algorithm. This theorem states that if certain derived conditions are met by composition, decomposition, and primitive solution operators, then a divide-and-conquer algorithm will satisfy a given problem specification. The strategy used to design the top level sorting algorithm is to choose a decomposition operator from the known primitives in the system, and to design the insertion algorithm to satisfy the derived specification for the composition operator. The insertion algorithm is designed by choosing a known composition operator and designing a custom decomposition operator (a simple conditional). This subproblem in the design is described in the second chart. In addition to the decomposition and composition operators, synthesis of a divide-and-conquer algorithm in Smith's formalism also requires specifying the predicate that tests the decomposability of an input, and designing or choosing an operator to compute the solution directly for an input that cannot be decomposed. If an input is decomposable into two subproblems, one is solved recursively and the other is solved by an auxiliary operator, which is just the identity function for both the top level and the insertion. The simple conditionals are constructed with the aid of a case analysis strategy.

State	Operator	Alternative	Rationale
Specification: Sort produces ordered list whose corresponding multiset is equal to multiset of input.			Given.
	Determine sort set and signature.		[Implementation] Recurrent type is `List(N)`
Sort set and signature from the algebra: ⟨{N, LIST(N)}, {Cons}⟩.		Sort set and signature from the algebra: ⟨{N, LIST(N)}, {Append}⟩.	[Selection] Other path gives other algorithms.

(continued)

State	Operator	Alternative	Rationale
	Determine component decomposition, auxiliary and composition problems.		[Goal] Satisfaction of divide-and-conquer correctness conditions. [Implementation] Type compatibility between decomposition and composition operators.
Subproblems with Id as auxiliary operator.			
	Choose well-founded ordering.		[Goal] Guarantee termination. [Implementation] Output domain of Length is **N**, which is a well-founded set.
Well-founded ordering on length.			
	Choose simple decomposition operator.	Choose simple composition operator.	[Selection] Other path gives other algorithms. [Implementation] FirstRest has domain = input type of specification, range = output type of specification
Simple decomposition is FirstRest.			
	Construct composition specification.		[Implementation] Types and conditions derived from previous choices and divide-and-conquer theorem.
Composition specification: insert into ordered list.			
	Construct composition operator.		[Implementation] See next chart (A).

(continued)

State	Operator	Alternative	Rationale
Divide-and-conquer program for Insert.			
	Specify guard.		[Goal] Ensure base case distinguished. [Implementation] Complement derived input condition for decomposition.
Guard tests for empty input.			
	Choose primitive (base case) operator.		[Goal] Sort for base case. [Implementation] Works on empty list.
Primitive operator is Id.			

The design steps for the insertion algorithm:

State	Operator	Alternative	Rationale
(A) INSERT:			
Specification: insert into ordered list.			
	Determine sort set and signature.		[Implementation] Recurrent type is List(N)
Sort set and signature from the algebra: ⟨{N, LIST(N)}, {Cons}⟩.			
	Specify component subproblems.		[Goal] Satisfaction of divide-and-conquer correctness conditions. [Implementation] Type compatibility between decomposition and composition operators.

(continued)

State	Operator	Alternative	Rationale
Subproblems with Id as auxiliary operator.			
	Choose well-founded ordering.		[Goal] Guarantee termination. [Implementation] Ordering maps $N \times LIST(N)$ to N, which is a well-founded set.
Well-founded ordering on length of list in input pair.			
	Choose simple composition operator.		[Implementation] Satisfies subproblem specifications.
Composition operator is Cons.			
	Construct decomposition operator.		[Implementation] Types and conditions derived from previous choice and divide-and-conquer theorem; case analysis strategy to synthesize conditionals.
Decomposition conditional.			
	Specify guard.		[Goal] Ensure base case distinguished. [Implementation] Complement of derived input condition to decomposition operator.
Guard tests for empty input list.			
	Choose primitive (base case) operator.		[Goal] Insert for base case input. [Implementation] Satisfies Insert specification if list in input pair is empty.
Primitive operator is Cons.			

Two notable features of this derivation also characterize the other derivations from Smith that we studied. The first is the procedural character of the design process. The user of the CYPRESS algorithm design system must still formalize the specification and domain knowledge, and make selections when several alternatives are available, but the design and deduction operators are applied in a predefined order. Limiting possible designs to a specialized class of algorithms, in this case simple divide-and-conquer algorithms, enables one to formalize the properties sufficiently so that such "design algorithms" are possible. The second feature is the heavy reliance of this derivation on a generalized deduction system, known as *antecedent derivation.* A derived antecedent is the information that must be added to a given set of premises to be able to conclude a desired goal expression. The design strategy used in this derivation makes frequent use of derived antecedents to test whether specifications are satisfied and to set up subproblem specifications.

2.4. Clark and Darlington

Clark and Darlington's derivation of insertion sort also appears within a collection of derivations, with the object of the presentation being to show how a "family" of algorithms can be derived from a single specification within a formal deductive system. In previous work, both authors presented more detailed descriptions of their languages and systems for program transformation; the article we studied uses a more informal version of the frameworks they developed to illustrate the close relationships between the four sorting algorithms they derived.

The derivation of insertion sort proceeds by expansion and reformulation of the definition of the sorting function. The expansion and reformulation is a consequence of execution of the specification on two symbolic inputs: a list with a single element, and a concatenation of two smaller lists. Together these cover the input domain of the sorting function, so the possible execution paths are divided into base and structure cases, leading to a recursive divide-and-conquer algorithm. For the base case, lemmas about sequences allow the simplification of one path into a direct computation of the output. For the structure case, the input is rewritten as a concatenation of two lists, and an intermediate result is introduced so that the conditions to be satisfied can be factored into weaker conditions to be satisfied separately. Because of knowledge about permutations and ordering, this new expression for the structure case can then be rewritten so that parts of the expression are satisfiable by recursively sorting sublists. The recursive call is introduced by applying the *fold* transformation (from the "fold-unfold" system of Burstall and Darlington [12]). The part of the expression not satisfied by the recursive calls is factored out as a specification for a new function that merges the two sorted sublists to produce a sorted permutation of the input. When one argument to this function is specialized to be a list of one element, this is a specification for insertion. Synthesis of the insertion algorithm is not described in the presentation.

State	Operator	Alternative	Rationale
Definition of sort as ordered permutation of input.			Given.
	Apply symbolic execution on base cases.		[Implementation] Inductive definition of list domain, lemmas about lists.
Sort(x) is x for empty and unit lists.			
	Substitute list pairs for input and output in sort definition.		[Implementation] Every list is a concatenation of two lists for nonbase cases.
Input and output split in two in rewritten sort definition.			
	Introduce intermediate result.		[Implementation] Transitivity and other lemmas for permutation
Intermediate result is structurally similar to input.		Intermediate result is structurally similar to output.	[Selection] Other path gives other algorithms.
	Fold.		[Goal] Introduce recursive call. [Implementation] Fold is forced by adding ordering conditions and alternate definition of ordering.
Recursion introduced, and residual is specification for merge.			

(continued)

State	Operator	Alternative	Rationale
	Specialize one of the arguments in merge to a unit list.	No specialization.	[Implementation] Lemmas about unit lists. [Selection] Other path gives merge sort.
Synthesis of insert left to reader.			

This synthesis omits much of the detail in the other presentations, but still illustrates two important concepts. The first is symbolic execution, using domain knowledge to reformulate and simplify the sort specification when the input is specialized to particular subsets of the input domain. Symbolic execution is a pervasive activity in human algorithm design, and can play a major role in an automatic algorithm design system as well [58]. The second concept is the use of "fold" transformations to introduce recursion. Not only is folding an important operator, but finding a state to which the fold operator can be applied is a goal that motivates much of the synthesis: Clark and Darlington add conditions to an expression in order to obtain the recurring pattern of evaluation required for the fold.

2.5. Composite

At the level of abstraction used in our analysis, there are few differences in the sets of major steps taken, though there are substantial differences in emphasis, level of detail, and justifications. We have combined all four derivations to show the major steps in the composite, Figure 2.1. The derivations all begin with the idea of solving a large problem by splitting it into more easily solvable smaller problems and combining the solutions.[2] At some point, all the derivations choose a decomposition that splits off a single element, and create the insertion to insert the element into the result of sorting the rest of the sequence. Sequences with zero or one element are already sorted and no further processing is needed. The syntheses of the insertion procedures (not present in Clark and Darlington's presentation) differ primarily because of the different target languages. Smith synthesizes a recursive divide-and-conquer algorithm and leaves the removal of the recursion and other refinements to a subsequent optimization phase. Both the Green and Barstow, and the Broy presentations eventually arrive at

[2] It is interesting that algorithm design textbooks, such as [52], do not present insertion sort this way.

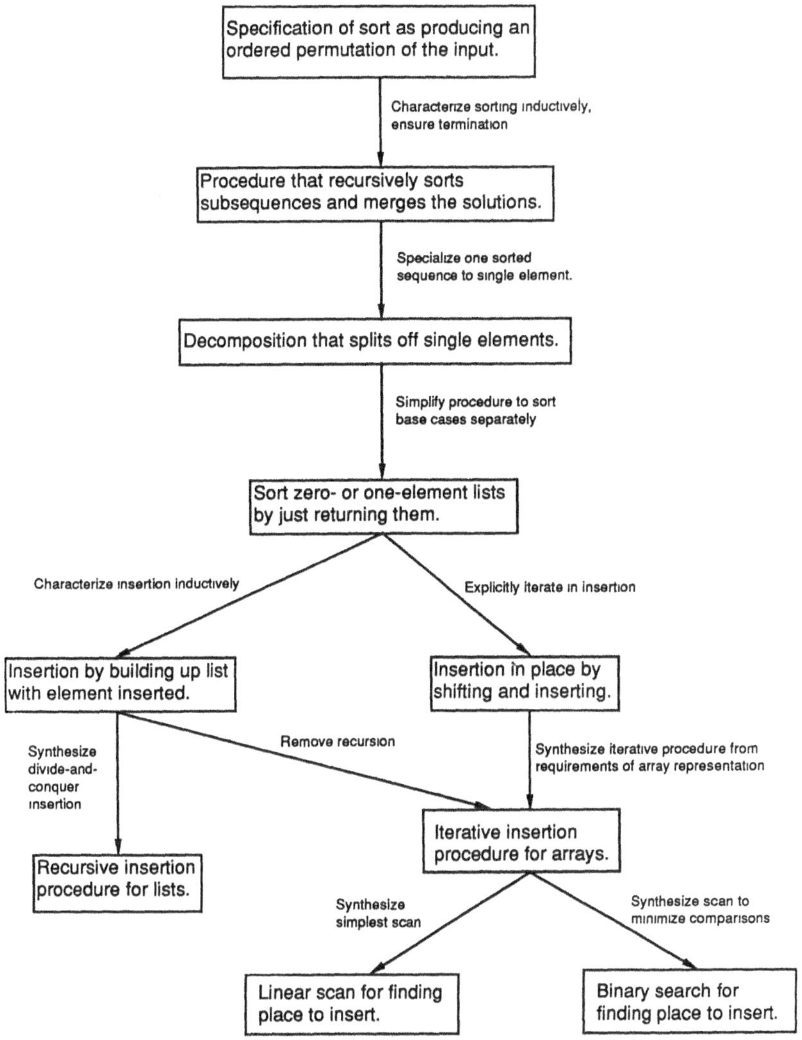

Figure 2.1. Composite for insertion sort derivations.

iterative insertion algorithms. Both presentations are significantly influenced by the choice of arrays as a representation for the input and output sequences, though Green and Barstow are more explicit than Broy in describing the last steps of the synthesis most directly affected by this choice. In considering refinements at that level, they confront issues of time and space complexity not addressed in the other presentations of the insertion sort algorithm.

CHAPTER 3

Quicksort

While insertion sort is widely used because it is so simple, it is possible to obtain much more efficient algorithms (when sorting large numbers of elements) by paying the price of a little extra complexity in algorithm design. For example, quicksort sorts by partitioning a sequence into two halves, sorting each half, and then combining the halves to produce a sorted sequence. The partitioning rearranges the sequence elements so that all elements to the left are less than or equal to all the elements to the right. The payoff is that the sorted subsequences can be merged by just concatenating them. Theoretically, partitioning around the median is necessary to achieve optimal performance (to balance sizes of the subproblems), but in practice an approximation to the median is good enough. Though quicksort has a potentially $O(n^2)$ performance, the average case performance is $O(n \log n)$. Hence it is the perferred sorting algorithm for large data sets in most cases.

The derivations compared here are from Green and Barstow [26], Broy [10], Smith [56], and Clark and Darlington [13].[1] Since the specifications for quicksort in these presentations were the same as for insertion sort, we move directly to a discussion of the results of the presentations. As can be seen from the following list, the final products of the quicksort derivations vary even more than for insertion sort. The presentation methods cover the full spectrum from English descriptions, to logical expressions, to recursive procedures, to optimized iterative code.

1. **Green and Barstow:** The article includes an English description of iterative quicksort split and join operations for sorting an array, illustrated by example.
2. **Broy:** The final algorithm is an iterative algorithm that manipulates stacks explicitly.

[1] Again the Clark and Darlington derivation updates a quicksort derivation using the same "fold-unfold" approach (but with generate-and-test) from [15]. Other derivations are by Hansson [30], Laaser [38], and Traugott [59].

```
funct quicksort = (array A) array:
 ⌈(nat[1 ... n] row var m X, var stack s1, var stack s2) :=
                              (A, push(empty, 1), push(empty, upb(A)));
   while ¬empty :: s do
   (var nat min, var nat max) := (top(s1), top(s2));
   p: do
        do if    min = max          then leave p
           elif X[min] ≤ X[max]   then min := min + 1
           ||   X[min] ≥ X[max]   then (X[min], X[max], max) := (X[max], X[min], max -̣ 1);
                                       leave
           fi
        od;
        do if    min = max          then leave p
           elif X[min] ≤ X[max]   then max := max - 1
           ||   X[min] ≥ X[max]   then (X[min], X[max], min) := (X[max], X[min], min + 1);
                                       leave
           fi
        od;
     od;
     if max - top(s1) > 1 then s2 := push(push(pop(s2), max - 1)), top(s2))
                          else s1 := pop(s1) fi
     if top(s2) - min > 1 then s1 := push(s1, min + 1)
                          else s2 := pop(s2) fi
   od;
   X;⌋
```

3. **Smith:** A recursive algorithm for lists is given by four nondeterministic conditionals, one for the top-level Qsort which calls Partition. In turn Partition calls Partition_Compose and Partition_Directly_Solve. The partition algorithm differs from the standard partitioning algorithm in that it is a divide-and-conquer algorithm and does not use a partitioning element (though one could be incorporated in a later optimization phase). If there are only two elements in the list, it produces two singleton lists, with the smaller element in the first list. Otherwise it builds up two lists by recursively partitioning off the rest of the list, and adding each element in turn to the appropriate sublist as determined by its value.

```
Qsort: x ≡
  if x = nil ∨ Rest: x = nil → Id: x  ||
     x ≠ nil ∧ Rest: x ≠ nil → Append ∘ (Qsort × Qsort) ∘ Partition: x
  fi

Partition: x ≡
  if Rest ∘ Rest: x = nil → Partition_Directly_Solve: x  ||
     Rest ∘ Rest: x ≠ nil → Partition_Compose ∘ (Id × Partition) ∘ FirstRest: x
  fi
```

```
Partition_Compose: ⟨b, ⟨z, z'⟩⟩ ≡
  if b ≤ Bag: z' → ⟨Cons: ⟨b, z⟩, z'⟩ ||
     b ≥ Bag: z → ⟨z, Cons: ⟨b, z'⟩⟩
  fi

Partition_Directly_Solve: x ≡
  if First: x ≤ First ∘ Rest: x → ⟨List ∘ First: x, List ∘ First ∘ Rest: x⟩ ||
     First: x ≥ First ∘ Rest: x → ⟨List ∘ First ∘ Rest: x, List ∘ First: x⟩
  fi
```

4. **Clark and Darlington:** The algorithm is given as two logical expressions. The second expression is a specification for partition, not refined in the article.

$$\text{sort}(x1\langle\,\rangle x2) \Leftarrow \text{sort}(y1)\langle\,\rangle\text{sort}(y2)\ \textbf{s.t.}\ \text{partition}(x1\langle\,\rangle x2, y1, y2)$$

$$\text{partition}(x, y1, y2) \Leftarrow \text{perm}(x, y1\langle\,\rangle y2) \wedge \forall u \forall v[u \in y1 \wedge v \in y2 \rightarrow u \leq v]$$

3.1. Green and Barstow

The first few high-level steps of Green and Barstow's quicksort synthesis, as in the syntheses by the other authors, are similar to those of insertion sort. The major difference is that quicksort uses a complex divide step that attempts to produce equal-sized subproblems, while insertion sort uses a simple divide step that splits off single element subproblems. Quicksort's divide step chooses a partitioning element, and then splits the input into a set of elements greater than this element, and a set of elements less than this element. To reduce the complexity of the implementation, the first element in the array is chosen as the dividing element rather than an approximation to the median. The merge doesn't have to do any work, since the outcome of solving the subproblems after partitioning leaves the elements in the correct places. The remainder of the synthesis focuses on details of the partition, such as where the partitioned elements are to be placed and the order in which the array is to be scanned while assigning elements to either set. A final optimization in the enumeration strategy, so that moves are only made when two elements need to be exchanged, results in the standard quicksort algorithm.

State	Operator	Alternative	Rationale
Specification: transfer elements to ordered collection.			Given.
	Choose top-level technique.		[Implementation] Assumed correctness conditions on split and join operation.

(*continued*)

State	Operator	Alternative	Rationale
Divide-and-conquer program.		Generate-and-test permutations.	
	Choose relative sizes of subproblems produced by split.		
Equal-sized split.		Singleton split.	[Selection] Choice is efficient, reducing number of comparisons.
	Decide how to split the sequence to be sorted.		
Partition by value around dividing element.		Partition by position.	[Selection] Other paths give other algorithms.
	Decide how to choose divider.		
Choose first element in sequence as divider.		Choose estimated median as divider.	[Selection] Selecting first element is simpler, though possibly less efficient.
	Choose technique for division.		[Goal] Get simplest division.
Generate elements from input and add to either one of two sets.			
	Decide where to store sets being created.		[Goal] Use space efficiently in sort.
In-place sort.		Not-in-place sort.	[Selection] In-place sort is feasible and uses space efficiently.
	Choose relative positions for input and two output sets.		[Implementation] Illustration by example.
Input set is between two output sets.		Two output sets are adjacent.	[Selection] Six relative position possibilities (narrowed to two for reasons of symmetry), efficiency in moves.

(continued)

State	Operator	Alternative	Rationale
	Change enumeration strategy.		[Goal] Efficiency of moves. [Implementation] Example showed some elements moved twice, intermediate moves can be eliminated.
Enumerate from both sides and exchange only when both need to be moved.			

As in the insertion sort synthesis, there is substantial comparison of alternative design paths throughout the quicksort synthesis. Decisions are made on the basis of time and space efficiency, though in some cases the desire for a simple implementation overrides runtime efficiency concerns. Also interesting in this synthesis is the use of examples to motivate the final optimizations. There is no formal proof of the fact that exchanging elements only when two elements need to be interchanged results in fewer moves than alternatives; informal reasoning supplemented with the example is taken as sufficient rationale. The use of examples in human-oriented presentations suggests that one might want to take examples as a source of knowledge in automated synthesis systems as well.

3.2. Broy

The path to quicksort in Broy's derivation shares the first decision, selecting an inductive basis for the algorithm, with the path to insertion sort. The paths diverge in the manner of decomposition of the input array into subproblems, where in quicksort the array is divided into one subarray with length one, and two others of nearly equal length containing the other elements. The singleton array is chosen nondeterministically, and the value of its element is used to partition the remaining elements to produce the other two arrays. The merge is just a concatenation of arrays. The derivation increases the efficiency of the algorithm by using an exchange primitive in the target language and dividing the partition into separate phases (leading to the scanning algorithm). Final refinements include recursion removal and open coding of simple subroutines.

State	Operator	Alternative	Rationale
Specification: Output array is ordered, has same elements as input.		Specification: Output array is ordered, corresponding multiset equal to multiset corresponding to input array.	Given.
	Apply induction principle.	Characterize array by number of inversions, reduce to 0 to sort.	[Selection] Other paths give other algorithms. [Implementation] Correctness given by proof by induction on length of array.
Algorithm sorts subarrays, then merges them.			
	Choose number of subarrays.		
Decomposition of array into three subarrays, one with length 1.		Decompose into two or more than three subarrays.	[Selection] Other paths give other algorithms.
	Decide how to choose singleton element.		
Singleton element chosen non-deterministically.		Choose element of maximum or minimum value, or by position in array.	
	Refine array split operation.		
Partition around chosen element by value.			
	Rewrite partition to exchange two elements at a time when needed.		
Recursive partition program.			

(continued)

State	Operator	Alternative	Rationale
	Optimize partition by splitting it into two phases.	Different methods for passing partition element as parameter.	[Implementation] Optimization possible after chosen element is maximum or minimum.
Rewritten recursive partition.			
	Remove recursion.		[Goal] Efficiency. [Implementation] Recursion removal by using stacks.
Iterative quicksort, with subroutines open-coded.			

One interesting feature of this derivation, which appears in some of the other presentations we studied, is the initial decision to have the partition element selected nondeterministically, to avoid making premature commitment to a particular implementation. Though not shown in the chart (apparently embedded in the recursion removal), a later step replaces the nondeterministic selection with a deterministic choice of the last element in the sequence, presumably made for reasons of convenience in the target language.

3.3. Smith

Smith's derivation of quicksort also appears within a collection of derivations for other divide-and-conquer algorithms, and thus follows the same approach as his derivation of insertion sort. The synthesis of the top level divide-and-conquer quicksort algorithm instantiates the divide-and-conquer scheme by choosing the simple operation of appending two lists for the composition, and the partition operation for the decomposition. The specifications for decomposition are satisfied by `FirstRest`, and a strategy for synthesizing conditionals is used to find simple programs for the composition and primitive operators. The complication in this synthesis comes in designing the partition algorithm. Our summary follows the chart for the top-level algorithm.

State	Operator	Alternative	Rationale
Specification: Sort produces ordered list whose corresponding multiset is equal to multiset of input.			Given.
	Choose simple composition operator.	Choose simple decomposition operator.	[Selection] Other path gives other algorithms.
Simple composition is Append.		Simple composition is Cons.	[Selection] Other path gives other algorithms.
	Choose auxiliary operator.		[Goal] Required for correct divide-and-conquer algorithm. [Implementation] Both inputs to Append are lists.
Auxiliary operator is recursive call to Qsort.			
	Construct decomposition specification.		[Implementation] Types and conditions derived from previous choices and divide-and-conquer theorem.
Decomposition specification: partition into two lists by value.			
	Construct decomposition to satisfy partition specification.		[Implementation] See (A).
Divide-and-conquer program to perform partition.			
	Derive guard.		[Goal] Ensure base case distinguished. [Implementation] Complement input condition to partition.

(*continued*)

State	Operator	Alternative	Rationale
Guard checks for singleton or null input.			
	Choose primitive (base case) operator.		[Goal] Sort for singleton or null input. [Implementation] Satisfies specification on base cases.
Primitive operator is Id.			

The synthesis of the partition algorithm is more complicated, because the input condition initially derived for the partition algorithm, `true`, is not strong enough. A first try fails to synthesize a primitive operator that can partition empty lists, so the input condition is revised to exclude the empty list as a legal input. A second try shows that singleton lists must be excluded as well. Only on the third try, after the input specification is modified to insure that the algorithm will only be invoked on lists with two or more elements, can an algorithm be derived. The derivation, then, is an illustration that for a certain class of incompleteness in specifications, a system implementing Smith's method for designing divide-and-conquer algorithms can overcome the incompleteness without additional direction from the user.

State	Operator	Alternative	Rationale
(A) PARTITION:			
Specification: Partition so that elements of first list are all less than elements of second list.			
	Choose simple decomposition operator for `Partition`.		

(*continued*)

State	Operator	Alternative	Rationale
Decomposition operator is `FirstRest`.			[Implementation] Matches input type of LIST(N).
	Choose well-founded ordering.		[Goal] Guarantee termination.
Ordering by length.			
	Choose auxiliary operator for `Partition`.		[Implementation] Input domain (N) of auxiliary function must be different from input domain of `Partition` (lists).
Auxiliary operator is `Id`.			
	Verify decomposition operator against specification.		[Goal] Ensure decomposition operator satisfies specification; find input condition to ensure decomposition always works when invoked. [Implementation] Nonempty input allows `FirstRest` to satisfy specification.
Decomposition works if list is nonempty.			
	Construct composition specification for partition.		[Implementation] Types and conditions derived from previous choices and divide-and-conquer theorem.
Composition specification: insert single element into one of two lists.			

(continued)

State	Operator	Alternative	Rationale
	Construct composition operator for partition.		[Implementation] Case analysis strategy to synthesize conditionals.
Conditional program for `Partition_Compose`.			
	Construct primitive operator to partition empty list.		
Specification for primitive operator unsatisfiable; derived input condition is `false`.			[Implementation] Can't partition empty list.
	Construct new input condition to partition.		[Goal] Ensure primitive operator never has to partition empty lists. [Implementation] Strengthen input condition using `else` branch of partition algorithm.
New input condition, so that `Partition` is not invoked on empty lists.			
	Verify decomposition operator with new input condition.		[Goal] Get new derived input condition to decomposition for use as guard.
`FirstRest` satisfies specification if input is not singleton list.			
	Construct new primitive operator to partition singleton lists.		
Specification is still unsatisfiable.			[Implementation] Can't partition singleton lists.

(*continued*)

State	Operator	Alternative	Rationale
	Construct new input condition to Partition.		[Goal] New input condition may allow satisfiable primitive operator specification. [Implementation] Combine current derived input conditions to FirstRest and Partition.
New input condition filters both empty and singleton lists.			
	Verify decomposition operator.		[Goal] Get new derived input condition. [Implementation] Derived antecedent after operator matching is input condition.
FirstRest satisfies specification if input has length >2.			
	Construct primitive operator to partition lists of length ≤2.		[Implementation] Case analysis synthesis strategy with rewritten input condition as guard.
Conditional program to perform Partition_Compose.			

In all the quicksort derivations studied, there seems to be a separation between the step that sets up the partition specification, and the next step, which creates an algorithm to satisfy the specification. Of the four presentations, this one pays the most attention to the construction of the specification, acknowledging both that it is possible to be systematic in constructing or augmenting specifications and that the process may involve iteration. In fact, this iteration makes our presentation format for the derivation as a sequence of steps somewhat misleading, as it should be a tree, whose branches are due to the necessity for backtracking. This backtracking shows up only a few times in the collection of derivations we studied, but there is evidence that further study of iteration in the derivation process will be needed for the development of derivation-based software tools [43].

3.4. Clark and Darlington

Clark and Darlington derive an expression for the top-level quicksort algorithm and a specification for the partition algorithm. As in insertion sort, the expression for quicksort is derived by symbolic execution of the sorting definition on two kinds of lists that together cover the input domain: the list with a single element, and the list composed of two smaller lists. Execution for the base case is the same as insertion sort. For the strucure case, again the synthesis begins by rewriting both the input and output lists as concatenations of smaller lists. However, for quicksort, the permutation condition chosen for factoring defines the relation between the intermediate result and the output, in contrast to insertion sort where the relation factored is between the intermediate result and the input. With some additional assumptions that both parts of the intermediate result are ordered, the condition that the output is ordered can be weakened to the condition that each element in the first subpart of the output is less than the elements in the second subpart of the output. This condition becomes the specification for the partition. When this condition is met, the merge just concatenates the two sorted sublists. With some additional equivalences and rearranging, the recursive calls can be introduced and the structure of the algorithm, though still fairly high-level, is in place.

State	Operator	Alternative	Rationale
Definition of sort as ordered permutation of input.			Given.
	Apply symbolic execution on base cases.		[Implementation] Inductive definition of list domain, lemmas about lists.
Sort(x) is x for empty and unit lists.			
	Substitute list pairs for input and output in sort definition.		[Implementation] Every list is a concatenation of two lists for nonbase cases.

(continued)

State	Operator	Alternative	Rationale
Input and output split in two in rewritten sort definition.			
	Introduce intermediate result.		[Implementation] Transitivity and other lemmas for permutation.
Intermediate result is structurally similar to output.		Intermediate result is structurally similar to input.	[Selection] Other path gives other algorithms.
	Fold.		[Goal] Introduce recursive call. [Implementation] Fold is forced by adding ordering conditions and alternate definition of ordering.
Recursion introduced, and residual is specification for partition.			

3.5. Composite

The composite chart (Figure 3.1) shows that, as in insertion sort, there is substantial overlap between the presentations, though significant variations in the later stages of the derivations. Abstracting from these variations, the essence of the design of quicksort can be captured in a few key steps. The derivations all begin with the inductive characterization of the sorting problem, leading to a divide-and-conquer algorithm. The base case for the algorithm is that a list with zero or one elements is already sorted. To treat the cases where the input must be split into smaller subproblems, there is a decision, implicit in some of the derivations, to focus the design effort on producing a division that makes the merge step simple. Another high-level design decision is to create a divide step that produces roughly equal-sized subproblems (a choice that may be motivated by reasoning about efficiency). The divide step is then specified to produce two sequences, where each element in the first sequence will be less than or equal to all the elements of the second sequence. The requirements of the divide step can be satisfied

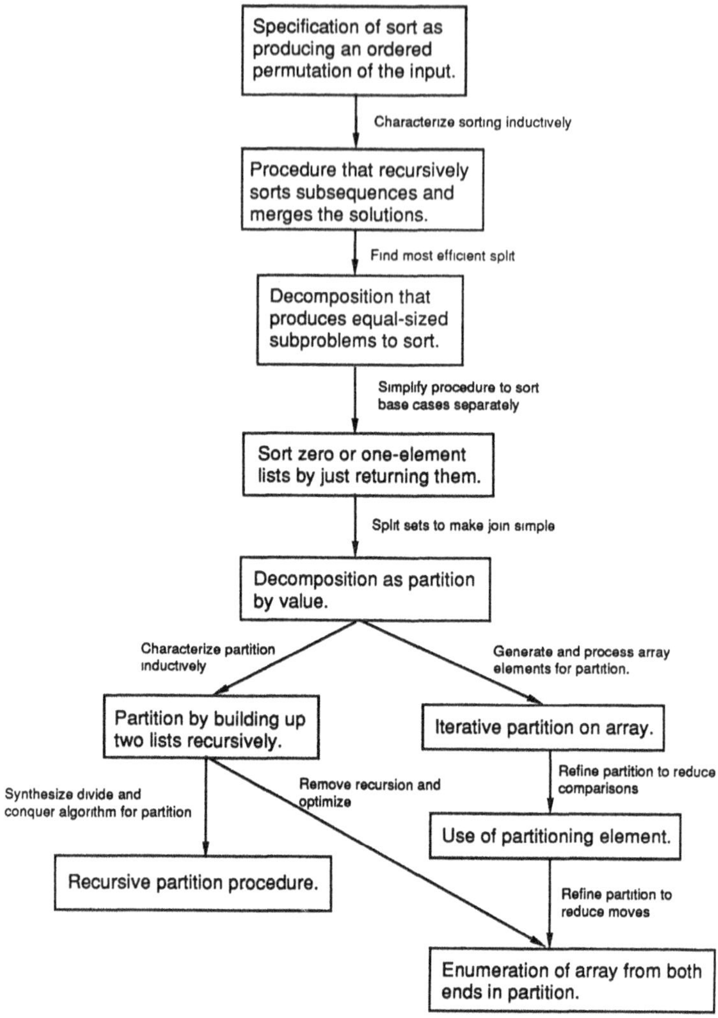

Figure 3.1. Composite for quicksort derivations.

by a divide-and-conquer algorithm or an iterative scan. A partitioning element may be used to minimize comparisons in the partition algorithm. Final refinements may focus on the method of selecting this partition element (approximating the median by sampling a few elements works best) and on the order in which elements of the list to be partitioned are scanned (in an in-place quicksort of an array, the array may be scanned from both ends towards the center simultaneously, exchanging two elements at a time when necessary).

CHAPTER 4

Cartesian Set Product

The Cartesian product of two sets is the set of all pairs of elements such that the first element of the pair is in one input set and the second element is in the other input set. We discuss derivations of algorithms to find Cartesian set products from Manna and Waldinger [42] and from Smith [53].

The input specifications for the problem are:

1. **Manna and Waldinger:** Compute the Cartesian product of two sets s and t:

   $\text{cart}(s, t) \Leftarrow \textbf{compute } \{(x, y) \mid x \in s \wedge y \in t\}$

2. **Smith:** Compute the Cartesian product of two sets defined as:

   $\text{cart-prod}: \langle x, x' \rangle = z \textbf{ such that } z = \{\langle a, b \rangle \mid a \in x \wedge b \in x'\}$
   $\textbf{where } \text{cart-prod}: \text{set}(\mathbf{N}) \times \text{set}(\mathbf{N}) \rightarrow \text{set}(\mathbf{N} \times \mathbf{N})$

Though the specifications differ in syntax, both express the same output condition. The fact that the inputs and outputs contain natural numbers is an incidental commitment of Smith's specification, and is not used in the derivation.

The outputs of the derivations are:

1. **Manna and Waldinger:** The output consists of two procedures in a Lisp-like language. Each is a single conditional. One calls the other to compute the Cartesian product of the head of its first argument and its second argument. For this derivation, the empty set, *head* and *tail*, union, and notations for singleton sets and pairs are considered primitive so that they don't need to be refined.

```
cart(s t) ⇐ if empty(s)
              then { }
              else carthead(s t) ∪ cart(tail(s) t)
```

```
carthead(s t)⇐if empty(t)
                then { }
                else {(head(s) head(t))} ∪ carthead(s tail(t))
```

2. **Smith:** The output is expressed as two recursive divide-and-conquer procedures in a functional programming language. The algorithmic primitives provided are *Set_Split* for decomposing a set into an element and the remainder of the set, *Add* to add an element to a set, *Id*, the identity operator, *Trans* (transpose) and *Union* (of sets).

```
CP: ⟨x, x'⟩ ≡ if x = { }
              then { }
              else Union ∘ (CP_aux × CP) ∘ Trans ∘ (Set_Split × Id2): ⟨x, x'⟩

CP_aux: ⟨a, x⟩ ≡ if x = { }
                 then { }
                 else Add ∘ (Id × CP_aux) ∘ Trans ∘ (Id2 × Set_Split): ⟨a, x⟩
```

The decomposition operator used for the top-level divide-and-conquer algorithm is Trans ∘ (Set_Split × Id2), where Trans transposes a tuple of typles, Set_Split splits off an element from a set returning the element and the remainder, and Id2 returns a 2-tuple containing two instances of its input. The effects of the decomposition are best illustrated with the example from p. 52 of [53]:

$$\text{Trans} \circ (\text{Set\_Split} \times \text{Id2}): \langle\{1, 2, 3\}, \{4, 5\}\rangle = \text{Trans}: \langle\langle 1, \{2, 3\}\rangle, \langle\{4, 5\}, \{4, 5\}\rangle\rangle$$
$$= \langle\langle 1, \{4, 5\}\rangle, \langle\{2, 3\}, \{4, 5\}\rangle\rangle.$$

The outputs of these derivations are at roughly equal levels of detail (though Manna and Waldinger explicitly specify a list representation of sets, while Smith is not committed to a particular representation). Excluding differences of syntax, the major difference between the two sets of procedures is in the **else** clauses of the conditionals. The outputs are operationally equivalent if one does some expansion at the top level and parameter substitution for the auxiliary function (the inputs to Smith's auxiliary function are a list element and a list, while for Manna and Waldinger the arguments are both lists, resulting in a slightly more complex function).

4.1. Manna and Waldinger

The derivation of an algorithm to find Cartesian set products is part of a collection of derivations to illustrate programming principles to be included in an automatic program synthesis system. This algorithm was synthesized automatically by their system, DEDALUS.

The problem is first decomposed into finding the Cartesian product of the singleton consisting of an element of the first set and all of the elements in the second set, and finding the product of the rest of the first set and the second set. Because the sets are represented as lists, *head* and *tail* are chosen to perform the decomposition, the *union* for composing subsolutions. Since the second subproblem is an instance of the specification at the top level, it can be solved recursively. The first subproblem is decomposed further into the union of a set containing a single pair (the singleton and one of the elements in the second set) and the Cartesian product of the singleton and the rest of the elements in the second set. The set containing the single pair is derived by the equality-elimination rule, which essentially simplifies set specifications. Because the goal of computing the other set is an instance of the goal to solve the entire subproblem, a new procedure `carthead` is introduced to solve the subproblem recursively. To terminate the recursion, the base case of an empty input is checked in both procedures, and the empty set is returned in these cases.

State	Operator	Alternative	Rationale
Specification: all pairs with one element in input set x and the other in set y.			Given.
	Decompose specification using head and tail of x (set represented as list).		[Goal] Get expression in a form where one part is a recursive call. [Implementation] Union-formation rule with knowledge about sets.
Specification is the union of two sets, the second of which is computed by a recursive call on the tail of x and y.			
	Decompose specification of first set using head and tail of y (also a list).		[Goal] Get specification of first set in a form where one part is directly computable. [Implementation] Equality elimination rule: forming a pair out of two elements is primitive; assume y nonempty.

(continued)

State	Operator	Alternative	Rationale
First expression replaced by union of a single pair (head of x and head of y) and the second subexpression.			
	Introduce new procedure `carthead`.		[Goal] Allow second subexpression to be computed by a recursive call to `carthead`. [Implementation] Assumption that sets are input and x is nonempty.
Output specification for `carthead` to compute the Cartesian product of the head of x and y.			
	Incorporate assumption of nonempty x as guard on recursive call in `carthead`.		[Goal] Ensure valid calls to `carthead` with conditional. [Implementation] Conditions for valid recursive call satisfied.
Final program.			

A number of principles, represented as rules about the specific application domain of sets and about programming in general, are applied in this derivation. The domain-specific rules are empty-set formation, union-formation, and equality elimination. The domain-independent rules are conditional-formation, recursion-formation, and procedure-formation. There are other principles included in Manna and Waldinger's derivation system, but it is not clear from the presentation where they were applied for this example. In most cases, the enabling conditions for transformation rules guarantee that a rule is relevant, but the relevant rules are ordered a priori, and sometimes rules have *strategic conditions* to prevent useless firings. For example, strategic conditions prevent conjunct reordering rules from firing in infinite sequence.

4.2. Smith

Smith's Cartesian set product algorithm is created by instantiating the divide-and-conquer program scheme, the same one used for the sorting and convex hull algorithms. The divide-and-conquer program scheme allows instantiation according to at least three design strategies. These strategies result from the freedoms allowed in choosing the composition, decomposition, and auxiliary operators, because the choice of one of these constrains the other choices. Here, Smith decides to use known decomposition and composition operators, and to derive the auxiliary operator from the specification set up from those choices. Though the choice of the particular decomposition operator (described on page 40) certainly requires some insight, Set_Split is considered as a basis of the decomposition operator because sets are relevant to the problem, though it has to be augmented for this problem since the input domain is 2-tuples of sets. The choice of composition operator, set union, is guided by the required input and output domains, as deduced from the already instantiated part of the divide-and-conquer scheme. The selected decomposition and composition operators set up a subproblem of finding the set of all pairs such that the first element in the pair is the first parameter and the second element is an element of the set passed as the second parameter. This specification is derived by reasoning backwards from the output condition, using a theorem about the divide-and-conquer algorithm and the output condition of the decomposition operator. The method of deduction used is the same antecedent derivation used for Smith's other derivations. The specification is satisfied by a divide-and-conquer algorithm similarly to the top level, but recursing along the second set rather than the first. The same strategy used for design of other divide-and-conquer algorithms is used here (see Smith's derivation of the top-level quicksort algorithm), but the presentation does not give details.

State	Operator	Alternative	Rationale
Specification: All pairs with one element in one input set and other in other set.			Given.
	Construct decomposition operator.		[Implementation] Input domain is 2-tuples of sets; `set_split` and `Id2` input domains are sets.

(continued)

State	Operator	Alternative	Rationale
Decomposition operator is Trans ∘ (Set_Split × Id2).			
	Choose output domain of auxiliary function.		[Goal] Domain must be known for choosing composition operator. [Implementation] Input domain to auxiliary function (**N** × **SET**(**N**)) is similar to input domain of top level function.
Output domain is **SET**(**N** × **N**).			
	Choose composition operator.		[Implementation] Set union has input/output types required in composition specification.
Composition is set union.			
	Create specification for auxiliary function.		[Implementation] Reasoning backward from instantiated correctness conditions for divide-and-conquer algorithm.
Specification for finding "product" of single element and set.			
	Synthesize algorithm for auxiliary function.		[Implementation] Divide-and-conquer design strategy
Divide-and-conquer algorithm for auxiliary function.			

Most of the designs of divide-and-conquer algorithms we have seen follow the strategy of designing composition operators given known decomposition operators, or of designing decomposition operators given known composition operators. Both of these strategies assume that both subproblems produced by the decomposition will be solved recursively, but Smith's formalism only requires a recursive solution to the one of the subproblems, leaving the decision about what to do with the other subproblem open. In most of his derivations, the other subproblem is either also recursively solved or else just passed along to the composition operator. Another possibility, explored in this derivation, is the design of a relatively complex auxiliary operator to apply to the second subproblem. This strategy opens up the possibility that procedure formation, a process requiring substantial "insight" in other design frameworks, might benefit from the constraints of Smith's design strategies.

4.3. Composite

The composite for the two derivations is shown in Figure 4.1. Abstracting from the details of each presentation, a key step in deriving this algorithm seems to be the reformulation of the specification as the union of two partial

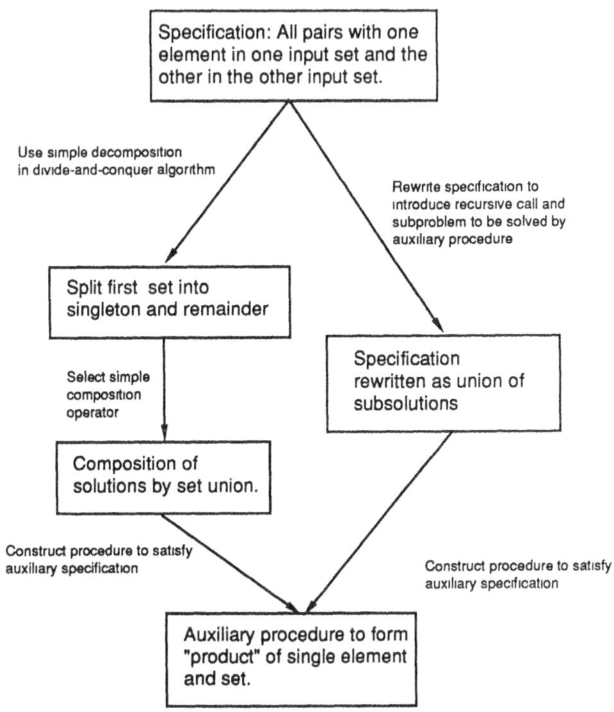

Figure 4.1. Composite for Cartesian set product derivations.

Cartesian products. The first is the Cartesian product of one element in the first set and the entire second set, and the second, computed recursively, is the product of the rest of the first set and the second set. Another key step is the introduction of an auxiliary procedure to compute the first partial product. This auxiliary procedure has a similar form to the top-level one, decomposing the expression for the output into two partial results agian, computing the first directly and the second by a recursive call to the auxiliary operator. Finally, domain knowledge is useful in finding the base case for both recursions, the empty inputs that can't be decomposed.

CHAPTER 5

Depth-First Search

In sequential algorithms, reachability in a directed graph is often determined by depth-first search. Starting at a given vertex, a depth-first spanning tree is constructed by recursively searching all successors. To prevent looping, each vertex is marked before visiting its successors, and a marked vertex is not searched again.

In this chapter we compare derivations by Broy and Pepper [11], Reif and Scherlis [50], Gerhart [22], and Barstow [7]. In the first three of these, depth-first search is developed as a preliminary step in derivations of more complex graph algorithms; Barstow, however, studies the reachability problem in isolation.

The four problem specifications follow:

1. **Broy and Pepper:** The specification is given as an equation defining the reflexive, transitive closure of a relation R for an element x:

$$A(R, x) \stackrel{\text{def}}{=} R^*(x)$$

with standard definitions for products of relations and reflexive, transitive closure:

$$RQ(x) \stackrel{\text{def}}{=} \bigcup_{y \in Q(x)} R(y)$$

$$R^* \stackrel{\text{def}}{=} \bigcup_{i \geq 0} R^i$$

2. **Reif and Scherlis:** The specification is given as a definition of the path relation:

$$\mathsf{path}(u, v) \Leftarrow (u = v \vee (\exists w \in \mathsf{Adj}(u))\mathsf{path}(w, v))$$

and an expression to compute the set of all vertices reachable from a given vertex u:

$$\{v \mid \mathsf{path}(u, v)\}$$

The recursive definition of the path relation, when interpreted in the natural way as a program, will not always terminate for graphs with cycles. The authors sketch a method of defining a fixed-point semantics for the specification in which the path relation is always defined; this semantics provides a basis for regarding the definition as a well-defined specification.

3. **Gerhart:** An asserted program in an imperative high-level language is given as the specification. Termination is ensured by maintaining the set of vertices already seen.

declare *Root* : **Integer**; **declare** *Seen*, *Unsure* : **SetSequence**;
assert $Root \in [1 \ldots n]$;
asserting $\mathsf{Tc1}(Root) = \mathsf{Tc}(Seen)$ and $Unsure \subseteq Seen \subseteq [1 \ldots n]$
  and $\mathsf{Rel}(Seen - Unsure) \subseteq Seen$
do $\neg(\mathsf{Rel}(Unsure) \subseteq Seen) \rightarrow$
    $\mathsf{Select}(Unsure, next, \neg(\mathsf{Rel1}(next) \subseteq Seen))$;
    $Seen := Seen \cup \mathsf{Rel1}(next)$;
    $Unsure := (Unsure \cup \mathsf{Rel1}(next)) - (\mathsf{seq}(next))$;
od
assert $\mathsf{Tc1}(Root) = Seen$

In this program, the transitive closure of the relation Rel1 on a particular vertex is denoted by Tc1, and its union over a set of vertices by Tc; thus $\mathsf{Tc1}(x)$ is the set of vertices reachable from the vertex x; $\mathsf{Tc}(X)$ is the set of all vertices reachable from the members of X. Similarly, the union over a set of vertices X of the relation Rel1 is denoted by $\mathsf{Rel}(X)$.

The **SetSequence** type, used to represent sets, provides both set and sequence operations on its elements. Since the sequence operations are not used until later design stages, the specification could be phrased entirely in terms of sets.

4. **Barstow:** An algorithm in an imperative high-level language is provided to find all vertices reachable in a directed graph from a given vertex. The marking of vertices to prevent looping is specified in terms of a mapping from integers, representing vertices, to strings, representing marks. The imperative constructs include the enumerators "for all" and "repeat until," without commitment to the order of enumeration. Thus the algorithm, while it does specify a method of avoiding nontermination due to cycles in the graph, does not specify a depth-first order of search.

DATA STRUCTURES

VERTICES	a collection of integers
SUCCESSORS	a mapping of integers to collections of integers
START	an integer
MARKS	a mapping of integers to {EXPLORED, BOUNDARY, UNEXPLORED}

```
ALGORITHM
  VERTICES ← input a list of integers;
  SUCCESSORS ← input an association list of
    ⟨integer, list of integers⟩ pairs;
  START ← input an integer;
  for all X in VERTICES:
    MARKS[X] ← UNEXPLORED;
    MARKS[START] ← BOUNDARY;
    repeat until MARKS⁻¹[BOUNDARY] is empty;
      X ← any element of MARKS⁻¹[BOUNDARY];
      MARKS[X] ← EXPLORED;
      for all Y in SUCCESSORS[X]:
        if MARKS[Y] = UNEXPLORED
              then MARKS[Y] ← BOUNDARY;
        output MARKS⁻¹[EXPLORED] as a list of integers.
```

All four specifications contain the basic idea of a relation and a vertex as inputs to the problem, but differ in the level of abstraction used. Broy and Pepper are fully abstract, specifying a relation R and a vertex x. Reif and Scherlis also specify a relation (Adj) and a vertex, but the relation is not an explicit parameter. Gerhart represents the relation as the pair of functions Rel, giving the successors of a set of vertices, and Rel1, giving the successors of a single vertex; vertices are represented as integers. Barstow is the most concrete, representing the relation as an association list and vertices as integers.

Similarly, Broy and Pepper, and Reif and Scherlis, specify the output abstractly, as a set. Gerhart specifies it as a **SetSequence**, a representation that provides both set and sequence operations. However, the sequence operations are not used in the specification. Barstow's specification is fully concrete, representing the output as a list of integers.

No method for ensuring termination in the presence of cycles in the graph is given in the specifications of Broy and Pepper or Reif and Scherlis. Gerhart and Barstow both give a method that depends on the maintenance of the set of vertices already seen. However, Barstow views this as a mapping from a vertex to one of three states, EXPLORED, BOUNDARY, and UNEXPLORED, which corresponds to the maintenance of three disjoint sets. Gerhart's method is to maintain the set of all vertices known to be in the transitive closure, *Seen*, and a subset of these, *Unsure*, whose successors have not yet been examined. Her set *Unsure* corresponds to Barstow's boundary vertices, but there is no set that corresponds to his explored vertices.

In general, the purely mathematical specification of Broy and Pepper is the most abstract. Reif and Scherlis give a specification in their language of programs, but the recursive definition of path will not always terminate if it is given the usual interpretation. (They describe briefly how their language can be given an unconventional semantics in which the function always terminates.) The use of the path relation is a feature that sets their specification apart from the others. Gerhart gives an iterative algorithm, with some commitment to the representations for sets and vertices. Barstow also gives an iterative algorithm, which is fully committed to representations for the input and output.

Next we consider the results of the four developments. In the case of the first three, these are not the last design stages presented by our sources, because they develop depth-first search as a preliminary stage in more complex algorithms. We have selected for presentation here the last stages before specialization to the larger problem begins. Thus these three "results" are more abstract than Barstow's, since his goal is to develop a Lisp implementation for the reachability algorithm.

1. **Broy and Pepper:** The result is still a set of equations, but the translation into programming language constructs is straightforward. The new definition B ensures termination by maintaining the set S of vertices already seen. This result is transformed further in their paper to obtain the Schorr-Waite algorithm (see Chapter 6).

$$A(R, x) = B(R, x, \{\ \})$$

$$B(R, x, S) = \begin{cases} S & \texttt{if } x \in S; \\ \bigcup_{y \in R(x)} B(R, y, S \cup \{x\}) & \texttt{otherwise} \end{cases}$$

2. **Reif and Scherlis:** The end product retains the abstract set data type and operations, but imperative constructs are used to manipulate a global array *visit*, with domain V, the set of all vertices of the graph. This ensures termination in the presence of cycles. This program is used, and further transformed, in other derivations in their paper.

```
path(u, v) ⇐ begin
                visit[V] ← false;
                v ∈ dfs(u)
             end

dfs(u) ⇐ begin
            visit[u] ← true;
            {u} ∪  ⋃         (if visit[w] then { } else dfs(w))
                 w∈Adj(u)
         end
```

3. **Gerhart:** A concrete implementation of general depth-first search is not derived, because the paper goes on to develop the Schorr-Waite garbage collection algorithm (see Chapter 6). We take the following program as the result for the purposes of this chapter because it is the last stage in Gerhart's derivation before specialization to the garbage collection problem is done.

```
assert Root ∈ [1 . . . n];
Marks, Finished, p, Stack := null, FALSE, Root, null;
asserting
  if Finished then Tc1(Root) = Marks
    else Tc1(Root) = Tc(Marks ∪ (Stack apr p))
        and Rel(Marks) ⊆ Marks ∪ (Stack apr p) ⊆ [1 . . . n]
        and NoDuplicateElements(Stack apr p)

do ¬Finished →
    if Rel1(p) ⊆ Marks ∪ (Stack apr p) →
        if Stack = null →
            Marks, Finished := Marks apr p, TRUE
        □ Stack ≠ null →
            Marks, p, Stack := Marks apr p, Last(Stack), OtherThanLast(Stack)
        fi
    □ ¬(Rel1(p) ⊆ Marks ∪ (Stack apr p)) →
        Select(Rel1(p), next, ¬(next ∈ Marks ∪ (Stack apr p)));
        p, Stack := next, (Stack apr p)
    fi
od
assert Tc1(Root) = Marks;
```

Although this program still makes use of abstract set operations, some refinement of data and control structures has been done. The **apr** operation appends an element to one end of a sequence. The set *Seen* of vertices that have been seen so far has been implemented as the union of the sequences *Marks* and *Stack*. The idea of the current vertex *p* has been introduced, and an efficient termination test implemented.

4. **Barstow:** Lisp code is produced. A paraphrase in the specification language is also given, which we show below. Sets are represented as lists of integers or arrays of Booleans, and the mapping *MARKS* is represented as a record containing two lists and an array. As a result of the data representation refinement, each statement in the main loop of the program can be executed in constant time.

```
VERTICES ← input a list of integers;
SUCCESSORS_input ← input an association list of
    ⟨integer, list of integers⟩ pairs;
SUCCESSORS ← create an array of size 100;
for all X in the list SUCCESSORS_input:
  SUCCESSORS[X:DOMAIN] ← X:RANGE
START ← input an integer;
MARKS:BOUNDARY ← create an empty list with header cell;
MARKS:UNEXPLORED ← create an array of size 100;
```

```
MARKS:EXPLORED ← create an empty list with header cell;
for all X in the list VERTICES:
  MARKS:UNEXPLORED[X] ← TRUE;
insert START at front of MARKS:BOUNDARY;
MARKS:UNEXPLORED[START] ← FALSE;
loop:
  if MARKS:BOUNDARY is the empty list then exit;
  X ← front element of MARKS:BOUNDARY;
  insert X at front of MARKS:EXPLORED;
  remove front element of MARKS:BOUNDARY;
  for all Y in the list SUCCESSORS[X]:
    if MARKS:UNEXPLORED[Y] then
      MARKS:UNEXPLORED[Y] ← FALSE;
      insert Y at front of MARKS:BOUNDARY;
repeat;
output MARKS:EXPLORED
```

The output of Reif and Scherlis's derivation is similar to Broy and Pepper's: both are recursive programs that compute a quantified union of recursive calls over the successors of a vertex. Reif and Scherlis are more concrete, representing the set of vertices already seen as a global array of Booleans, while Broy and Pepper have made no commitment to a representation. The use of global state is a device that permits Reif and Scherlis to delay commitment to an order of evaluation for the set union operations in their algorithm. Thus, the depth-first ordering is not explicit in their result, but a left-to-right execution would result in a depth-first search. In Gerhart's and Barstow's programs, the depth-first ordering is imposed by the implementation of the set of boundary vertices as a stack. Gerhart has made no further commitments to set representation, but Barstow's program uses fully concrete Lisp data structures.

5.1. Broy and Pepper

The depth-first search derivation of Broy and Pepper is carried out entirely by mathematical reasoning. It is a preliminary stage in a development of low-level executable code for the Schorr–Waite graph marking algorithm.

The first stage in the presentation is the formulation of equations that define the standard mathematical notions of relations on finite sets, the product of relations, and reflexive transitive closure; the specification is then given as a function A mapping a relation R and an element x to the transitive closure $R^*(x)$. A recurrence relation for R^* is found by rewriting the definitions, and this in turn is used to find a recurrence for A, giving the prototype of a recursive program that would not terminate in the presence of cycles in the graph. In the next step, it is observed that in order to ensure termination, one may remove from the graph every edge that has already been traversed. This idea is formalized by defining a relation $[R_x]$, which is R modified to remove outgoing edges of x. The original specification of A is then shown to be equivalent to a recursive equation in terms of $[R_x]$. But since an algorithm based on the new definition would have to modify the original graph, a further improvement is made by generalizing A to a function B having an extra parameter, the set of all elements already seen. This final equation directly corresponds to a depth-first search algorithm, operating on the abstract data types of sets and relations.

State	Operator	Alternative	Rationale
Equational definition of reflexive, transitive closure of a relation.			[Goal] Generality and formality.
	Find recurrence.		[Implementation] Equational reasoning (rewriting).
Recursive equation.			
	Introduce modified relation.		[Goal] Obtain terminating definition. [Implementation] Mathematical induction; lemmas about closures of relations.
Recursive equation that modifies the relation on recursive calls.			
	Add auxiliary parameter.		[Goal] Obtain terminating definition in which relation is not modified. [Implementation] Lemmas about modified relation and closures of relations.
Recursive equation maintaining set of elements known to be in closure. Depth-first ordering.			

In this presentation, the justification for each step is given in the form of fairly detailed proofs. The emphasis is on a clear and concise mathematical presentation, with some intuitive appeal to the goals that guide the process.

5.2. Reif and Scherlis

This derivation is a preliminary to the development of several complex graph algorithms, including strong-connectivity and biconnectivity. Program transformation techniques (as opposed to mathematical reasoning) are used throughout.

The first stage is a recursive definition of the path relation in a directed graph. This definition, when interpreted in the usual way as a sequential program, does not terminate in the presence of cycles. The reachability problem is stated as the formation of the set of all vertices v such that there is a path from a given vertex u to v. The first step is one of specialization: the recursive path definition is specialized to the set expression, yielding a recursive definition for the reachability problem. This definition is more efficient than the specification, which evaluates path for every possible pairing of u with another vertex. Except for language, the definition is identical to the second stage of the Broy and Pepper derivation. Viewed as a conventional program, it does not always terminate. Termination is obtained in the next step, in which the finite closure transformation is used to introduce a global Boolean array that represents the set of vertices already seen. Finite closure is a technique that can be applied to some recursive functions defined over finite sets, resulting in new definitions in which redundant recursive calls are eliminated.

State	Operator	Alternative	Rationale
Definition in terms of recursive relation path.			
	Specialize path relation to set expression.	Finite closure.	[Goal] Efficiency. [Selection] Specialization is more difficult after finite closure.
Recursive procedure for set expression.			
	Finite closure: introduce global Boolean array.		[Goal] Obtain conventional terminating program.

(continued)

State	Operator	Alternative	Rationale
Recursive procedure that maintains set of elements known to be in closure, represented as global Boolean array. Depth-first ordering.			

Reif and Scherlis deal with the same issues explored by Broy and Pepper's derivation, but in a briefer, less formal style. The problem of ensuring termination for this algorithm is viewed as an instance of a more general problem, to be solved using the general finite closure transformation. As a result, the justification for the derivation is not mathematical reasoning about relations, but a proof (only a sketch of the proof is given) that the finite closure transformation is applicable.

5.3. Gerhart

Gerhart's development is a prelude to a derivation of a low-level implementation of the Schorr–Waite graph marking algorithm. It is unique among the sources we have studied, in that design stages are programs with assertions suitable for Hoare-style reasoning, and design steps modify the assertions and programs togehter.

The specification is an iteration that continues as long as some successor of a vertex in the set Unsure is still unseen. This expensive loop guard is removed by the first step, yielding a loop that processes a single element by either discarding it or adding its successors to the set to be processed. In the next step, operations on the set Unsure are restricted to list operations; this imposes the depth-first order on the search. For simplicity's sake, our presentation next compresses several of Gerhart's steps into one, which she identifies as an important derivation pattern: the introduction of a new variable which, after a few steps, replaces an old one. In this case, the set Seen, which is a superset of Unsure, is eliminated; the new variable Marks is the difference of Seen and Unsure. The next major step is again a variable introduction/elimination, which results in the elimination of the set Unsure, in favor of a stack; redundant computations of the next element of Unsure are eliminated by assigning it to a new variable.

State	Operator	Alternative	Rationale
Iterative program that maintains Seen = set of elements known to be in closure; Unsure ⊆ Seen, = set of elements whose successors are not in Seen. Computes successors of all elements of Unsure at each iteration.			
	Move computation from loop guard to loop body.		[Goal] Efficiency.
At each step, remove and compute successors of a single element of Unsure. Terminate when Unsure = { }.			
	Treat a set as a list.		[Implementation] No duplicate elements in Unsure.
List operations for element selection and maintenance of Unsure set; depth-first ordering imposed.			
	Eliminate a set S, and a set $R \subseteq S$, introducing disjoint sets $S-R$ and R.		[Implementation] Unsure ⊆ Seen.
The set Seen eliminated in favor of disjoint sets Unsure and Marks.			
	Introduce Boolean variable for loop guard		[Goal] Separate final and intermediate conditions in assertions.

(*continued*)

State	Operator	Alternative	Rationale
Loop tests and sets Boolean variable.			
	Eliminate a set treated as a list, introducing a sequence; eliminate the head of the list, introducing a new variable.		
Unsure eliminated in favor of the sequence Stack and the vertex p.			

Many features of the final algorithm are already present in the initial state of this presentation; thus many of the design steps are refinements of data or control structure rather than major algorithmic changes. For example, in the specification, the set of vertices Seen is maintained to ensure termination, and its subset Unsure provides control over the iteration. This basic structure is retained throughout, although the sets are evantually represented as lists. Moreover, knowledge of the desired result of the development is clearly used in the choice of the SetSequence data type to represent the sets from the very start.

The main feature that sets off this design sequence from the others is the view of the set of "boundary" vertices as a subset of the vertices already seen. In the recursive formulations used by Broy and Pepper, and Reif and Scherlis, this boundary set is not explicitly represented; in Barstow, disjoint sets are implied by the mapping MARKS. The author does not discuss the reasons for this choice. However, this scheme is clearly implied by the recursive versions: in these, a vertex is added to the set of vertices already seen, and at the same time it is (implicitly) added to the parameter stack. The shift to disjoint sets emerges as efficient implementations are sought for the sets. As a result, the later stages of Gerhart's presentation approach the early stages of Barstow's.

5.4. Barstow

Barstow's treatment of depth-first search is part of an investigation into the automatic implementation of abstract algorithms. The object of the development is to produce a *program* for reachability in a directed graph, given an abstract *algorithm*; depth-first search is not a specific goal (breadth-first would implement the abstract algorithm as well), but is a by-product of the rules that specify how to implement abstract operations and data types.

We suppress some of the details in the interests of readability. For example, the development of a representation for the SUCCESSORS mapping takes the following steps: a mapping is represented as an explicit mapping, i.e., a collection of pairs; an explicit mapping is stored in a single structure; an association table is selected for the structure; the table is represented as an array; the array is represented by a Lisp array; and as a result of six more (unspecified) steps, the elements of the array are chosen to be Lisp lists. Our presentation compresses all of this into one step.

Since the focus is on implementation, rather than algorithm design, the first stage of the design already contains the idea of marking vertices already seen, and the input and output are specified in terms of concrete data types. The first few steps concern the choice of representation for the SUCCESSORS mapping that represents the directed graph internally. The input has been specified to be an association list, and the abstract form of the algorithm merely assigns this list to the mapping. Once the choice is made to represent the mapping as an array of lists, implementations are developed for the computation of the image of an element under the map and the conversion of the input to the map representation.

In the abstract algorithm, the state of the search is maintained as a mapping from vertices to one of three states, EXPLORED, BOUNDARY, or UNEXPLORED. In the next few steps, a representation for the mapping is chosen and the corresponding operations are implemented. Representations are then developed for the BOUNDARY and UNEXPLORED sets. The BOUNDARY set is represented as a Lisp list, which imposes a stack discipline on the associated operations, leading to a depth-first order of search. A bit map representation is used for the UNEXPLORED set.

State	Operator	Alternative	Rationale
Represent successors mapping as an association list.			
	Represent mapping as array of lists.		[Implementation] The domain of the mapping is a fixed range of integers.
successors mapping represented as array of lists.			
	Implement map lookup as array access.		

(*continued*)

State	Operator	Alternative	Rationale
successors(x) implemented as array access.			
	Replace conversion of input to mapping, by conversion of input to array of lists.		
Convert input by: for all vertices in input, store successors in array.			
	Implement "for all" as total enumeration of elements.		
Convert by looping over input, storing successors of each element in array. Main body of algorithm maintains a mapping marks.			
	Represent mapping as a list of sets.		[Goal] Reduce cost of frequent computation of inverse mapping.
marks represented as a list of disjoint sets marks:boundary, marks: explored, and marks: unexplored.			
	Implement computation of inverse mapping as cons cell access.		
marks^{-1} implemented as cons cell access.			
	Implement modification of mapping as removal from one set and addition to another.		

(continued)

State	Operator	Alternative	Rationale
Assignment to marks implemented as modifications of appropriate sets.			
	Represent set as a list.		
boundary set represented as list.			
	Implement removal of arbitrary set element as removal of car of list; implement addition of element to set as cons.		[Implementation] boundary set is represented as a list.
Stack discipline for marks:boundary. Depth-first ordering.			
	Represent set as Boolean array.		[Implementation] The only operations on the unexplored set are membership test, add, and delete.
	Implement removal from set as assignment to array element.		
unexplored set represented as Boolean array.			
	(Other steps omitted by Barstow.)		
Iterative Lisp program.			

An unusual feature of this development is the use of a mapping as the initial abstraction for maintaining state. In most presentations, set membership is treated as the primary notion, but in Barstow, sets emerge only as a representation is sought for the mapping. It is not clear whether this use of a mapping would be a natural outcome of an algorithm development from an abstract specification.

Another noteworthy feature is that the representation for an abstract data type is fully developed before the corresponding operations on the type are implemented. In all the other depth-first search derivations, any particular design stage is self-consistent, and even executable, given a powerful enough interpreter. The design steps may therefore be considered atomic. In contrast, here there are design stages in which the operations on a data object are not consistent with its representation; further design steps are required before an executable (high-level) program is obtained.

5.5. Composite

In sketching the composite (Figure 5.1), we have added some steps (represented by dotted lines and omitted some details, particularly data representation choices from Barstow. The resulting composite emphasizes three main issues in the development: first, the method of ensuring termination in the presence of cycles; second, the conversion of the set S (all vertices known to be in the transitive closure) and its subset U (vertices whose successors may not yet be in S) to the disjoint sets $S - U$ and U; and third, the use of a list to represent the set U, which imposes a stack discipline and, as a result, a depth-first order on the search. We omit discussion of another important issue, the selection of a representation for S, because of the lack of detail in our sources.

The derivations of Reif and Scherlis, and Broy and Pepper, fit into the upper half of the composite, where the termination issue is dealt with. Although there are two distinct states reached in this part of the design space, the distinction between them is not a fundamental difference in the algorithm, but a difference in concreteness of data representation. The finite closure transformation of Reif and Scherlis leads directly to a very concrete representation of the set S as a Boolean array. Because Broy and Pepper's derivation is a prelude to the Schorr–Waite algorithm, this implementation decision is delayed until after the depth-first search algorithm is fully developed. It is interesting that, although the two presentations begin with quite different specifications, they pass through the same state before diverging again. Moreover, the specialization step of Reif and Scherlis is essentially a way of combining function definitions (some of them recursive) to arrive at a new recursive function definition; this corresponds closely to seeking a recursive equation as in Broy and Pepper. These correspondences suggest that the specifications may not be as different as they look, in terms of the information contained in them.

Below the point where termination has been ensured, the composite covers the presentations of Gerhart and Barstow. We have inserted a recursion removal step to cover the gap between Broy and Pepper's result and Gerhart's specification. Here, the depth-first ordering is made explicit by the selection of a list representation for U, and the sets S and U are

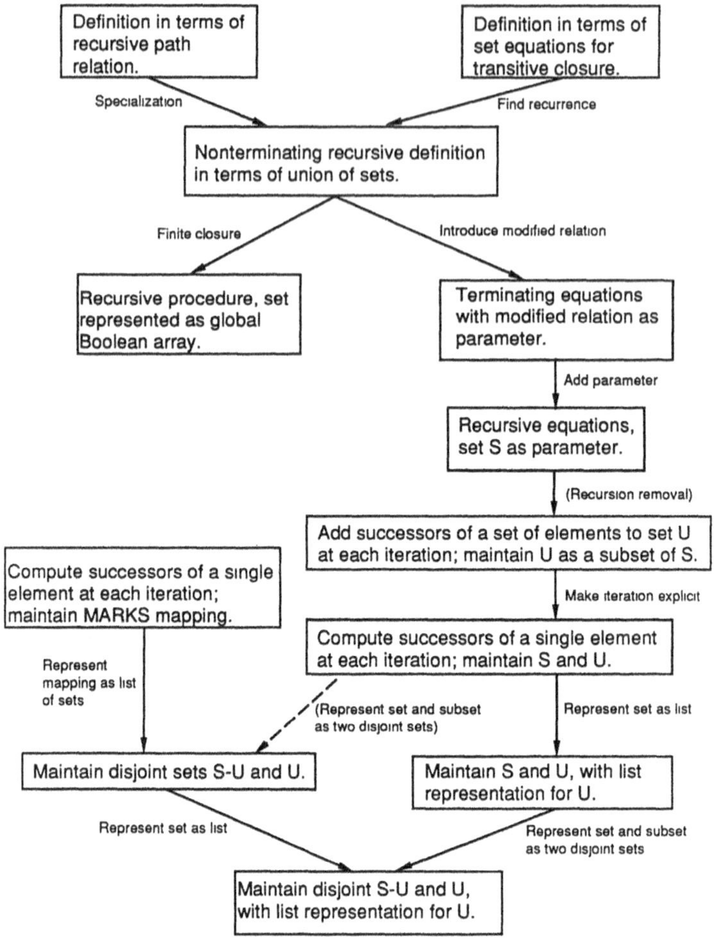

Figure 5.1. Composite for depth-first search derivations.

replaced by $S - U$ and U. It is evident that these two steps can be done in any order; we have inserted another step to show a design sequence that appears possible, though it is not shown in any of our sources.

We have omitted some steps of Barstow's synthesis from the composite, since there are no analogous steps in our other sources. We view these as further details of implementation that, if included, would extend the composite derivation graph into the realm of concrete implementations.

CHAPTER 6

Schorr–Waite Graph Marking

The algorithm of Schorr and Waite [51] is a specialized method for computing the reflexive transitive closure of a relation. The special case is Lisp garbage collection; thus the relation is represented by a directed graph with a maximum outdegree of two. The basic idea of the method is to traverse the graph depth-first, but avoid the maintenance of a parameter stack. Instead of a stack, the algorithm maintains a record of the path from a vertex to the root by "turning around" the pointers that represent the graph to point to a vertex's parent.

In this chapter we analyze three derivations of an elegant variant of this algorithm, developed by David Gries [27]. Gries's own presentation takes a form closer to that of proof than derivation, so we have not included it. Derivations by Gerhart [22] and Griffiths [29] consider only the variant of Gries. Broy and Pepper [11] derive several versions, including that of Gries. We have omitted derivations by Dershowitz [16] and Jonkers [32], for reasons of both space and clarity.

To clarify the comparison of the derivations, we have analyzed the early stages of [22] and [11], which derive general depth-first search algorithms, in Chapter 5. Here we consider the later stages, beginning with depth-first search and ending with Schorr and Waite's algorithm.

1. **Broy and Pepper:** The following group of equations for the transitive closure of a relation R, which can be viewed as a partial specification for this problem, is derived from a more abstract specification of reachability as described in Chapter 5.

$$A(R, x) \stackrel{\text{def}}{=} R^*(x)$$

$$A(R, x) = B(R, x, \{\ \})$$

$$B(R, x, S) = \begin{cases} S & \texttt{if } x \in S; \\ \bigcup_{y \in R(x)} B(R, y, S \cup \{x\}) & \texttt{otherwise} \end{cases}$$

The rest of the specification is an informal statement of the Lisp garbage collection problem. The graphs are binary, and can be represented by a pair of functions *ls* (left successor) and *rs* (right successor). These are extended to total functions by adding pairs $\langle x, x\rangle$ to the relation R for each x in its domain.

2. **Gerhart:** Again, this "specification" is derived from a more abstract specification (see Chapter 5.) It is an asserted program, with abstract data types and auxiliary definitions, that specifies an iterative depth-first search for the transitive closure Tc1(x) of a relation Rel1.

```
assert Root ∈ [1 . . . n];
Marks, Finished, p, Stack :=null, FALSE, Root, null;
asserting
  if Finished then Tc1(Root) = Marks
    else Tc1(Root) = Tc(Marks ∪ (Stack apr p))
         and Rel(Marks) ⊆ Marks ∪ (Stack apr p) ⊆ [1 . . . n]
         and NoDuplicateElements(Stack apr p)

do ¬Finished →
    if Rel1(p) ⊆ Marks ∪ (Stack apr p) →
        if Stack = null →
             Marks, Finished := Marks apr p, TRUE
        □ Stack ≠ null →
             Marks, p, Stack := Marks apr p, Last(Stack), OtherThanLast(Stack)
        fi
    □ ¬(Rel1(p) ⊆ Marks ∪ (Stack apr p)) →
        Select(Rel1(p), next, ¬(next ∈ Marks ∪ (Stack apr p)));
        p, Stack := next, (Stack apr p)
    fi
od
assert Tc1(Root) = Marks;
```

In this program, the transitive closure of the relation Rel1 on a particular node is denoted by Tc1, and its union over a set of nodes by Tc; thus Tc1(x) is the set of nodes reachable from the node x; Tc(X) is the set of all nodes reachable from the members of X. Similarly, Rel(X) dnotes the union over a set of nodes X of the relation Rel1. The apr operation pushes an element onto the stack, the Last operation accesses the top of the stack, and OtherThanLast pops the stack.

3. **Griffiths:** The starting point of the derivation is a recursive destructive marking algorithm for depth-first search. The graph is represented by an array, indexed by node numbers, containing four fields: a value, a mark, a left successor, and a right successor. Successor fields are 0 if no successor exists, and a marked dummy node 0 with no successors is added to the input graph.

```
procedure mark(x);
  if m(x) = 0
  then m(x) := 1;
       mark(left(x));
       mark(right(x));
  fi
endproc
```

All the specifications contain the idea of maintaining a set of nodes that have already been seen. Griffiths, however, is already committed to representing the set by a mark field in the node. The other two specifications use the notion of set (S in Broy and Pepper, *Marks* in Gerhart) without commitment to a particular representation. All three specifications also include the idea of depth-first search. Because Gerhart works in an iterative language from the very beginning, the parameter stack needed for the depth-first search is maintained explicitly as a sequence. In the other specifications the use of recursion allows the stack to remain implicit.

Although the Schorr-Waite algorithm is optimized for space usage, none of the specifications expresses any limitation on available space. This failure to express complexity constraints is typical, given the state of the art in specification languages. The algorithm reuses storage in a way that would not be employed in the absence of space limitations. Thus it might be an interesting case study for research into methods for specifying complexity properties.

The results of the derivations are especially interesting to compare in this case, since all of our sources aimed at deriving a variant of the algorithm developed and proved by Gries in [27].

1. **Broy and Pepper:** Five variants of the algorithm are derived, including that of Gries. They differ only in the details of recursion removal. All are derived from a parameterless recursive procedure that manipulates a global set s and another global data structure called an *extended graph.* This data type has the rotation (*rot*) and exchange (*ex*) operations and a current vertex (v) needed to carry out the basic Schorr-Waite idea. The processing for the current vertex v falls under one of four cases:
   (a) On the first visit, v is added to the set s, and a rotation makes the left child of v the current vertex for a recursive call.
   (b) When the processing of the left child is complete, another rotation makes the right child of v the current vertex for a recursive call.
   (c) After the processing of the right child, another rotation restores the left and right child pointers of v, making the parent of v the current vertex as the last step of the procedure.
   (d) If v is already in the set s, an exchange terminates the recursion.

   Although the programs are expressed in terms of abstract data types, for example, sets, in another section of this paper Broy and Pepper show how these can be implemented in terms of arrays.

The five solutions are of two types; we examine one of each type. The first of these depends on the introduction of a stack, a method that is generally applicable and therefore might be automatically found by a synthesis system. The other is Gries's variant of the program, which is one of a class of variants that arise when an auxiliary set is introduced. For each version we show an asserted recursive procedure definition from the original paper, and the iterative solution we have obtained by instantiating Broy and Pepper's recursion removal rule.

In the first version, recursion is eliminated by introducing a stack a of Booleans. The use of a stack is, of course, a standard recursion removal tactic; Booleans are sufficient since there are only two control points to be distinguished. The $s \oplus v$ notation denotes the adjoin operation (add v to s).

```
proc P ≡ ( ):
  if v ∈ s then ex
         else s ⊕ v;
              rot; a push 0; call P; {top a = 0} pop a;
              rot; a push 1; call P; {top a = 1} pop a; rot fi
```

The assertions in braces establish the form needed for recursion removal:

```
P:
  if v ∈ s then ex; goto R
         else s ⊕ v; rot; a push 0; goto P fi
R:
  if empty a then skip
           else if top a = 0 then pop a; rot; a push 1; goto P
                            else pop a; rot; goto R fi fi
```

The other result we examine is the variant of Gries, if implementations for the abstract data types are added. Here, recursion is eliminated by introducing auxiliary sets t and s', in such a way that the original set s is equal to their union. An implementation that represents set membership as a two-bit array element for each node can implement the set operations of this version by incrementing this variable on each visit, as in Gries's program.

Introducing a dummy root vertex *vroot* allows an efficient termination test; processing is complete when the auxiliary pointer *aux* of the extended graph returns to *vroot*.

```
proc P ≡ ( ):
  if v ∈ s' ∪ t then ex
           else s' ⊕ v; rot; call P; {v ∉ t ∧ v ∈ s'}
                s' ⊖ v; t ⊕ v; rot; call P; {v ∈ t ∧ v ∉ s'}s' ⊕ v; rot fi
```

After recursion removal:

```
P:
  if v ∈ s' ∪ t then ex; goto R
                else s' ⊕ v; rot; goto P fi
R:
  if aux = vroot then skip
               else if v ∉ t ∧ v ∈ s' then s' ⊖ v; t ⊕ v; rot; goto P
                                        else s' ⊕ v; rot; goto R fi fi
```

2. **Gerhart:** The result is the program of Gries, with assertions needed for a correctness proof. The assertions are given in abbreviated form.

```
assert Rootsinit and LRInit and mInit;
p, q := Root, VirtualRoot;
asserting some Stack(
  if p = VirtualRoot then MarksInvariant(p, Stack, m, TRUE)
    else MarksInvariant(p, Stack, m, FALSE)
        and LRStackInv(Stack, m, L, r)
        and LRotherinv(p, Stack)
  )
do p ≠ VirtualRoot →
    m[p] := m[p] + 1;
    if m[p] = 3 or m[L[p]] = 0 →
        p, L[p], R[p], q := L[p], R[p], q, p,
    □ m[p] ≠ 3 and m[L[p]] ≠ 0 →
        L[p], R[p], q := R[p], q, L[p]
    fi
od
assert mRootFinal(m) and LRFinal(L, R)
```

3. **Griffiths:** A close approximation to the variant of Gries is produced. The only difference is in the termination test:

```
x := root; q := 0;
do m(root) ≠ 3 →
    m(x) := m(x) + 1;
    if m(x) = 3 cor m(l(x)) = 0 →
        x, l(x), r(x), q := l(x), r(x), q, x
    □ m(x) ≠ 3 cand m(l(x)) ≠ 0 →
        l(x), r(x), q := r(x), q, l(x)
    fi
od
```

The results of Gerhart and Griffiths closely resemble each other if the assertions of Gerhart are ignored. Broy and Pepper's result is both more and less abstract than this. It is more abstract in its use of set operations, but less abstract in its use of `gotos` instead of structured iterations. Its relation to the two other programs is shown in the composite diagram.

6.1. Broy and Pepper

Broy and Pepper's derivation uses algebraic reasoning and program transformations to provide an explanation and a verification by construction of the algorithm. The use of an applicative language for the first stages of design leads to straightforward proofs and transformation operators. Because selective updating of data structures is not expressible in an applicative language, the pointer exchanges that are of central importance to the algorithm are developed using abstract data types. The final few steps remove the inefficiency inherent in the applicative program.

The initial state is a recursive depth-first search algorithm, expressed in mathematical equations, for general directed graphs. The first step specializes this algorithm to the Lisp garbage collection problem, in which no vertex has more than two successors. The specialization replaces the quantified union of recursive calls over all successors of a vertex with a binary union over the left and right successors. The two recursive calls are then transformed to a nested recursion, which is both more efficient and easier to transform. The next few steps establish the main features of the algorithm. First, the equations are translated to a typed functional language, and abstract data type definitions are given for graphs and sets. Then the essential idea of Schorr and Waite is introduced, via the definition of a new abstract data type, the *extended graph*, an extension of the type of graphs that includes the pointer manipulation operators *rotation* and *exchange*. The algebraic properties of the two data types are used to justify the use of the extended graph in place of the original graph. The resulting program contains some redundant pointer manipulations, which are eliminated in the next two steps. At this point, the essential ideas of the algorithm are complete, but the advantages of the method can be realized only by an iterative program that modifies variables. Transformation of the function definition to a recursive procedure that modifies global variables prepares for recursion removal. Broy and Pepper use a general recursion removal rule that depends on the development and proof of suitable assertions about the procedure. They examine five variants of the algorithm that arise from the use of different assertions. We summarize two of these. The first, labeled (A) in the summary chart, is obtained by the generally applicable method of introducing an explicit stack. The stack size may be as large as the number of vertices in the graph. The other solution, labeled (B) in the chart, leads to Gries's program. A virtual root is introduced in order to get an efficient termination condition, and a pair of sets, s', and t, replaces the original set s of vertices already seen. Gries's program results from the implementation of the two sets together as an array of two-bit "counting marks."

State	Operator	Alternative	Rationale
Depth-first search defined in terms of the union over all successors of a node.			Given.
	Replace union over all successors by union over left and right successors.		[Goal] Specialize to Lisp garbage collection problem. [Implementation] Maximum outdegree of a node in the graph is 2.
	Change multiple recursion to nested recursion.		[Goal] Efficiency and ease of program transformation. [Implementation] Mathematical proof; definition introduction.
Depth-first search in terms of nested recursive calls for left successor and right successor.			
	Translation from mathematics to typed functional language with abstract data types.		[Goal] Representation independence; ease of program transformation; ease of design automation.
Applicative program operating on **bigraph** data type.			
	Introduce new data type by adding distinguished nodes and extra operations to **bigraphs**.		[Goal] Prepare to eliminate parameter stack by using Schorr–Waite insight, i.e., manipulate graph to eliminate need for stack. [Implementation] Equational reasoning; computation induction and mathematical induction.

(continued)

State	Operator	Alternative	Rationale
Applicative program operating on **extended graphs** (**bigraphs** with **rotate** and **exchange** operations).			
	Introduce new function definition.		[Goal] Eliminate redundant **exchange** operations. [Implementation] exchange (exchange(e)) = e.
Program that calls the previous version, then does an **exchange** on the resulting graph.			
	Series of folding and unfolding steps.	Justify next state by inductive proof.	[Goal] Obtain recursive function definition. [Implementation] Lemmas about **exchange**.
Recursive definition without redundant exchanges.			
	Replace parameters by global variables; fold-unfold.		[Goal] Prepare for recursion removal.
Parameterless recursive procedure.			
	(A) Introduce stack a of Booleans, with assertions.	See (B).	[Goal] Remove protocol (control) stack. [Selection] Stack works in all cases. [Implementation] Booleans are sufficient because there are only two recursive calls to be distinguished.
Asserted procedure with stack operations.			
	Recursion removal.		[Goal] Efficiency.

(continued)

State	Operator	Alternative	Rationale
Iterative program operating on sets and stacks.			
	(B) Introduce virtual root in graph.	Introduce a counter.	[Goal] Simple termination test. [Selection space efficiency.
	Introduce two sets s' and t, whose union is the set of nodes seen.	(1) Introduce set t of vertices whose left successors have been searched. (2) Introduce set t of vertices whose left successors, but not right successors, have been searched. (3) Introduce two disjoint sets s' and t, whose union is the set of nodes seen.	[Goal] Remove protocol (control) stack. [Selection] Obtain Gries's variant of the algorithm.
Asserted procedure maintaining sets s' and t.			
	Recursion removal.		[Goal] Efficiency.
Iterative program operating on sets; an abstract form of Gries's variant.			

In this derivation, care has been taken to separate the various issues that contribute to the final algorithm: the specialization of depth-first search to binary graphs, the introduction of pointer manipulations to eliminate the need for the parameter stack, and the introduction of set manipulations to eliminate the need for the control stack. As a result, the goals that motivate each step are clear. Moreover, control and data abstractions are retained in the design as long as possible, until the basic ideas have been developed. The use of abstract data types to model destructive storage reuse operations is central to the development, allowing the algorithm to be justified by algebraic reasoning.

Although mechanization is not a major issue, there are mechanical methods tacitly or explicitly underlying some of the steps: the fold-unfold rules ([12]) used to implement some of the optimizations, and the recursion removal rule are two examples.

6.2. Gerhart

Gerhart's derivation combines program verification ideas with a derivation approach. The design states are asserted programs, and as the design evolves, the assertions are transformed along with the program. Once the derivation is complete, stages are presented to a mechanical verifier for proof, with some human assistance in organizing the proof.

The initial design is an iterative depth-first search algorithm for general directed graphs, which maintains a set of vertices already searched, and an explicit parameter stack of vertices still to be searched. First, the algorithm is specialized to the Lisp garbage collection problem by introducing a binary relation in place of the general one, and separating the processing of the left and right successors of a vertex into distinct cases. At this stage, a set membership test is still used to determine the state of a vertex. The next step introduces an array of counting marks to represent the set membership, and the program determines a vertex's state by testing its count. The rest of the derivation is concerned with applying the Schorr-Waite idea. A virtual root replaces the Boolean variable that flags termination. Then the stack is eliminated by introducing two arrays of left and right successors, which an auxiliary vertex pointer. The rotation and exchange operations, which are related to the stack by assertions, manipulate the arrays to provide backward pointers in the graph. The assertions permit the elimination of the stack. The resulting program is a loop containing three cases. Two case merging steps produce Gries's program.

State	Operator	Alternative	Rationale
Iterative depth-first search, operating on the set of all successors of a node.			
	Replace general relation by binary relation; case analysis on possible successors of a vertex.		[Implementation] Maximum outdegree in the graph is two.
Separate operations for left and right successors. Vertex test: test for membership of successors in a union of sets.			

(continued)

State	Operator	Alternative	Rationale
	Represent set membership of vertices by array of counting marks.		[Goal] Supply details of representation. [Implementation] Variable introduction and elimination.
Vertex test: test value of counting mark. Termination test: test flag, which is set when stack empty.			
	Replace Boolean flag with special value for current vertex, which is set when stack empty.		[Implementation] Variable introduction and elimination.
Termination test: test for current vertex = virtual root.			
	Embed parameter stack in two arrays and a vertex pointer.		[Implementation] Variable introduction and elimination.
Parameter stack replaced by auxiliary pointer and rotation, exchange operations on arrays. Loop has three distinct cases.			
	Merge cases.		[Goal] Obtain Gries's variant. [Implementation] Two of three cases are syntactically identical.
Loop with two distinct cases; one has two subcases.			

(continued)

State	Operator	Alternative	Rationale
	Merge cases.		[Goal] Obtain Gries's variant. [Implementation] One subcase is equivalent to another top-level case.
Gries's program; two simple cases.			

Gerhart's derivation makes extensive use of an informal technique we have called *variable introduction and elimination*, which she identifies as an important derivation pattern. This technique introduces a new variable that has no effect on the input-output behavior of the program, with assertions that describe its relation to the old variable. Then the new variable is used instead of the variable to be replaced, and finally the assignments to the old variable are deleted, and references to it in assertions are existentially quantified. This process eventually shifts all the information about the abstractions of the specification out of the program statements into the assertions. This technique is applied both to obtain concrete implementations of abstract data types, and to replace the stack maintenance by pointer manipulations. The assertions of the final program are quite complicated and difficult to prove; the paper focuses on the use of derivation techniques to make a mechanized proof tractable.

6.3. Griffiths

The derivation of Griffiths is an exercise in program explanation and an illustration of programming techniques. Since education, not mechanization, is the goal, the presentation is quite informal.

The initial design is a recursive depth-first search procedure for binary graphs that maintains a global Boolean array. Since an iterative program is wanted, some additional mechanism is needed to maintain the control state for a vertex; this is introduced in the first step as an array of counting marks instead of Booleans. Next, the recursion is eliminated to obtain a design for an iterative program. The design is incomplete, because it calls an undefined function to obtain the predecessor of a vertex. This function is then implemented by the pointer updates and accesses of the Schorr-Waite method. Gries's program is obtained by merging identical cases.

State	Operator	Alternative	Rationale
Recursive depth-first search in binary graph, using global Boolean array for making nodes.			Given.
	Replace Boolean marks by counting marks.		[Goal] Prepare for recursion removal.
Recursive depth-first search using counting marks.			
	Recursion removal.		
Iterative program using (undefined) predecessor function to maintain control context.			
	Replace call to predecessor function by left pointer access; update left and right pointers appropriately.	Replace call by right pointer access; with appropriate updates.	[Goal] Implement predecessor function. [Selection] This version leads to Gries's variant. [Implementation] Auxiliary variable introduction.
Predecessor function implemented by altering pointers.			
	Merge cases.		[Goal] Obtain Gries's varient.
Gries's program: two simple cases.			

A feature of this presentation that sets it apart from the others is the use of the predecessor function as an abstraction that motivates and guides the introduction of the pointer manipulations. The other aspects of the algorithm are treated at the relatively concrete level of arrays and assignments.

6.4. Composite

To help in our discussion of the composite (shown in Figure 6.1), we will follow Broy and Pepper in distinguishing two major aspects of depth-first search: the maintenance of a control context, and the maintenance of a parameter context. In a recursive procedure definition, such as the specification of Griffiths, both contexts are usually thought of as residing in a single stack, which contains the return address and the local binding environment for each invocation of the procedure. In this case, the return address is the control context, and the binding environment is the parameter context. In the case of Gerhart, the specification is iterative, not recursive. The parameter context is mantained as an explicit stack, but the control

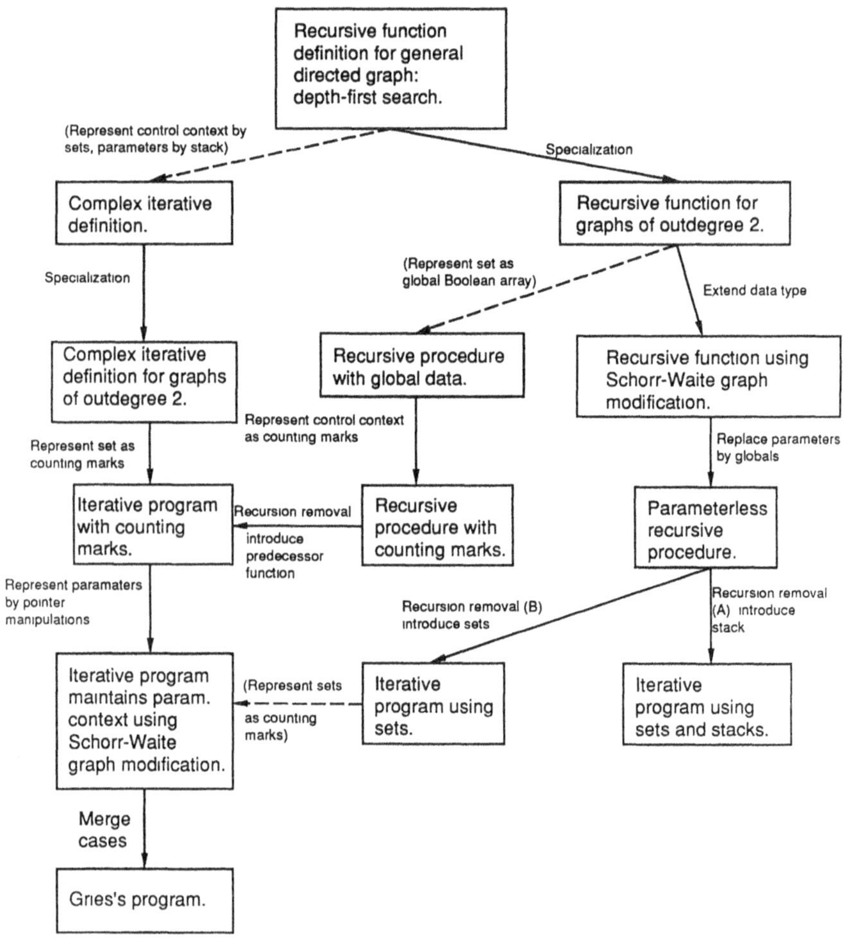

Figure 6.1. Composite for Schorr-Waite graph-marking derivations.

context is determined by the complex subset test in the outer **if** statement. The specification of Broy and Pepper is a recursive equation rather than a program definition, but it too can be viewed as maintaining control and parameter context on a stack. Broy and Pepper note that the insight of Schorr and Waite applies to the efficient maintenance of the parameter context only, and that variants of the algorithm arise from different strategies for representing the control context.

It is evident from the three derivations and their composite that the two issues are quite separate. The two aspects of the development commute, that is, either the parameter context or the control context can be refined first, and the derivations differ in this ordering.

The left-hand and middle paths represent Gerhart's and Griffiths's derivations, respectively, in which the control context is treated first. We have added two hypothetical steps (represented as dotted lines) to indicate how the specifications of Gerhart and Griffiths might derive from a common source. Gerhart works entirely in terms of iteration; consequently, in the initial design stages, set membership tests determine the control context, and the tests are then more efficiently implemented by counting marks. Griffiths takes a different approach, in which the set abstract data type does not appear. He takes as given a recursive procedure manipulating a global Boolean array, which in our composite is derived as a representation of a set. The Booleans are then changed to counting marks as a preparation for recursion removal. After recursion removal, the two derivations converge, roughly speaking. The pointer manipulations that are fundamental to the algorithm are introduced, in Gerhart, to replace the parameter stack; in Griffiths, to implement the predecessor function.

The right-hand path represents the derivation of Broy and Pepper. Here the order is inverted: the pointer manipulations that maintain the parameter context are introduced before the recursion removal step that makes the control context explicit. The five variants of the algorithm that they derive, only two of which we examine here, differ only in the methods of efficiently maintaining the control context. We have inserted another step to show how Broy and Pepper's variant (B) leads to Gries's program, given the appropriate representation choice for sets.

To emphasize the common features of the derivations, we have glossed over the issue of an efficient termination test in our composite. Griffiths uses, without commenting on the alternatives, a termination test that is not discussed in either of the other two papers. It is rather easy, however, to justify his test in terms of the recursion removal rule used in Broy and Pepper. A full treatment of this issue would result in a proliferation of paths through the parts of the derivation graph that deal with the representation of the control context.

CHAPTER 7

N-Queens

The problem of placing some number of queens, usually eight, on a chessboard so that no queen can capture another is a classic example of a problem that lends itself to solution by a backtracking algorithm. In this chapter, we compare derivations of the algorithm for this problem by Balzer [4], Smith [55], and Wirth [61].

The input specifications are:

1. **Balzer:** The specification is expressed in the Gist language [40]. Balzer's presentation does not give a complete specification, and states that additional declarations need to define objects and actions in the problem such as the chessboard, the eight queens, and the rules for queen-capture. The two constraints should be interpreted as *prohibiting* the situation they describe from occurring.

```
PROGRAMS:
  QUEENS: [LAMBDA (QUEEN-SET)
                  (LOCAL (BOARD-POSITION)
                         (FOR QUEEN IN-SET QUEEN-SET DO
                              (DETERMINE BOARD-POSITION FROM
                                         (CHESS-BOARD BOARD-POSITION))
                              (ASSERT (PIECE-ON-BOARD QUEEN
                                                      BOARD-POSITION]
  CONSTRAINT: TWO-PIECES-CAN'T-OCCUPY-SAME-SQUARE
              PATTERN: (AND (PIECE-ON-BOARD PIECE#1 BOARD-POSITION)
                            (PIECE-ON-BOARD PIECE#2 BOARD-POSITION))
              PATTERN-VARIABLES: (PIECE#1 PIECE#2 BOARD-POSITION)

  CONSTRAINT: QUEEN-CAN'T-CAPTURE-ANOTHER-QUEEN
              PATTERN: (AND (PIECE-ON-BOARD QUEEN#1 BOARD-POSITION#1)
                            (PIECE-ON-BOARD QUEEN#2 BOARD-POSITION#2)
                            (QUEEN-CAPTURE BOARD-POSITION#1
                                           BOARD-POSITION#2))
              PATTERN-VARIABLES: (QUEEN#1 QUEEN#2 BOARD-POSITION#1
                                  BOARD-POSITION#2)
```

2. **Smith:** A generator of solutions to the problem is the basis for the specification. The generator enumerates all arrays M, where $M[i]$ is the row number of the queen placed in column i. The first line in the condition on the array ensures that no two queens occupy the same row, while the last two lines in the condition ensure that no queens are on the same diagonal.

   generator Queens(k:N): **array**$[1 \ldots k]$ **of** $[1 \ldots k]$
     assume $1 \leq k$;
     enumerate $\{M \in$ **array**$[1 \ldots k]$ **of** $[1 \ldots k] \mid$
   $\forall a, b \in [1 \ldots k][a \neq b \Rightarrow M(a) \neq M(b)] \wedge$
   $\forall a, b \in [1 \ldots k][a \neq b \Rightarrow M(a)+a \neq M(b)+b] \wedge$
   $\forall a, b \in [1 \ldots k][a \neq b \Rightarrow M(a)-a \neq M(b)-b]\}$ **do**
       return(M)
     end

3. **Wirth:** The specification is given in informal English:

   > Given are an 8×8 chessboard and eight queens which are hostile to each other. Find a position for each queen (a configuration) such that no queen may be taken by any other queen (i.e., such that every row, column, and diagonal contains at most one queen).

Only Wirth's specification does not contain a precommitment to a form of an algorithm; both Balzer's and Smith's specifications contain a generate-and-test algorithm that theoretically could be executed to solve the problem. The specifications vary further in degree of abstraction; Wirth's and Balzer's are more abstract, since they are written in the terminology of the problem domain, rather than that of programming. They are also more modular; for example, the rules defining queen-capture are stated separately from the problem constraint that no queen-captures take place. In Smith's specification, these facts are intermingled. Smith's specification also commits one to a specific array representation of the board and particular methods for checking for captures. The constraints that no two queens occupy the same column is already implicitly enforced by the chosen representation. In setting up a specification this way, Smith has already imported some of the knowledge that must be used later in Balzer's derivation. Another difference is that Smith's specification requires finding all valid ways to place the queens, while Balzer's specifies finding just one. Wirth first only requires finding any solution, and later generalizes his algorithm to find them all.

We now compare the results of each derivation:

1. **Balzer:** The final state for this derivation is a backtracking program, including detailed incremental updating code for maintaining the sets

in case of successful and unsuccessful queen placements. However, the program could be refined further: set operations have been used freely, and no data structures have been selected to implement these operations efficiently. The program maintains sets for the remaining rows and columns allowed for placing queens. At each stage, a row and column is selected from these sets, and the program is recursively called with the remaining queens, rows, and columns to see if the assignment is part of a complete solution. If not, the row and column are added back to the pool of candidates, and another choice is made for the queen being placed. When no more queens are left to place, a solution has been found. The asterisks in some of the commands denote some of the information used in the derivation: that the commands are nondeterministic, but that the nondeterminism is not essential to the correctness of the algorithm.

```
(LAMBDA (QUEEN-SET REMAINING-ROWS REMAINING-COLUMNS
                   REMAINING-NE-DIAGONALS REMAINING-SE-DIAGONALS
  (LOCAL (QUEEN BOARD-POSITION BOARD-POSITION#2 BOARD-POSITION#1 QUEEN#2
         QUEEN#1 PIECE#2 PIECE#1)
        (TERMINATION-TEST: (IF (EMPTY QUEEN-SET)
                               THEN (EXIT SUCCESSFUL)))
        (REMOVE* QUEEN FROM QUEEN-SET)
        (FOR ROW IN-SET REMAINING-ROWS THEREIS
             (FOR COLUMN INSET REMAINING-COLUMNS
              WHEN
             (PROGN (DETERMINE* NE-DIAGONAL FROM
                         (CORRESPONDING-NE-DIAGONAL ROW COLUMN))
                 (DETERMINE* SE-DIAGONAL FROM
                         (CORRESPONDING-SE-DIAGONAL ROW COLUMN))
                 (APAND (IN-SET? NE-DIAGONAL REMAINING-NE-DIAGONALS)
                     (IN-SET? SE-DIAGONAL REMAINING-SE-DIAGONALS)
                     ))
             THEREIS
             (MAINTENANCE-ACTIONS: (REMOVE* ROW FROM REMAINING-ROWS)
                         (REMOVE* COLUMN FROM
                             REMAINING-COLUMNS)
                         (REMOVE* NE-DIAGONAL FROM
                             REMAINING-NE-DIAGONALS)
                         (REMOVE* SE-DIAGONAL FROM
                             REMAINING-SE-DIAGONALS))
             (DETERMINE* BOARD-POSITION FROM
                  (CORRESPONDING-BOARD-POSITION BOARD-POSITION
                                       ROW COLUMN))
             (ASSERT (PIECE-ON-BOARD QUEEN BOARD-POSITION))
             (IF (SUCCESSFUL (RECURSIVE-CALL: (QUEENS QUEEN-SET
                                              REMAINING-ROWS
                                       REMAINING-COLUMNS
```

```
                              REMAINING-NE-DIAGONALS
                              REMAINING-SE-DIAGONALS)
          THEN (LOOP-RETURN CHOICE-ACCEPTED)
          ELSE (UNDO-ACTIONS: (DENY (PIECE-ON-BOARD QUEEN
                                   BOARD-POSITION))
                    (MAINTENANCE-ACTIONS: (ADD ROW TO
                                 REMAINING-ROWS)
                                  (ADD COLUMN TO
                              REMAINING-COLUMNS)
                                  (ADD-NE-DIAGONAL TO
                            REMAINING-NE-DIAGONALS)
                                  (ADD-SE-DIAGONAL TO
                            REMAINING-SE-DIAGONALS))
               (LOOP-RETURN FORCE-ANOTHER-CHOICE))
       THEN (EXIT SUCCESSFUL)
       ELSE (UNDO-ACTIONS: (ADD QUEEN TO QUEEN-SET))
          (EXIT UNSUCCESSFUL)
```

2. **Smith:** The code is given in terms of two generators: the first ensures that all possible solutions are enumerated, and the second does the actual backtracking. The second generator (which can be simplified further) assumes that all the queens placed so far are safe, and then computes all the possibilities for safely placing a queen in the next column, invoking itself once for each possible position with updated array and indices. The same array representation has been retained from the specification, and has been used as the basis for testing for acceptable positions for queens. Set operations are used in the program, though the incremental updating procedures for maintaining the sets efficiently are not specified.

```
generator Queens(k:N):array[1 ... k] of [1 ... k]
  assume 1 ≤ k;
  if true then Queens_aux(k, 0, φ) endif
end

generator Queens_aux(k:N, i:N, m:array[1 ... k] of [1 ... k]):array[1 ... k] of [1 ... k]
  assume 1 ≤ k ∧ i ≤ k ∧
         ∀a, b ∈ [1 ... i][a ≠ b ⇒ m(a) ≠ m(b)] ∧
         ∀a, b ∈ [1 ... i][a ≠ b ⇒ n(a) + a ≠ m(b) + b] ∧
         ∀a, b ∈ [1 ... i][a ≠ b ⇒ m(a) − a ≠ m(b) − b];
  if i = k
   then if true then return(m) endif
   else enumerate l ∈ [1 ... k] do
          if l ∉ Valset(i, m) ∧
            l − i − 1 ∉ Diffset(i, m) ∧
            l + i + 1 ∉ Sumset(i, m)
            then Queens_aux(k, i + 1, m ⊕ [i + 1 ↦ l]) endif
   endif
end
```

where

$$\mathsf{Valset}(i, m) \equiv \{m(d) \mid d \in [1 \ldots i]\}$$

$$\mathsf{Sumset}(i, m) \equiv \{m(d) + d \mid d \in [1 \ldots i]\}$$

$$\mathsf{Diffset}(i, m) \equiv \{m(d) - d \mid d \in [1 \ldots i]\}$$

3. **Wirth:** The final program we present here is the iterative version given by Wirth; he also briefly discusses a recursive program to solve the same problem. It is written in a variant of ALGOL-60 (one which we now recognize as being very PASCAL-like), so its control constructs and data representations are those found in conventional programming languages. On each iteration, the program attempts to find a safe square on which to place the next queen. If a safe square is found, the program continues; otherwise, the last queen placed is removed, and a new position for it is tried. This iteration continues until either all the queens have been placed, indicating success, or there are no more positions to try, indicating failure.

   We have expanded some one-line procedures inline to shorten the program.

```
j := 1; i := 0;
repeat
  repeat i := i + 1; testsquare
  until safe ∨ (i = 8);
  if safe then
  begin setqueen; x[j] := j + 1; i := 0
  end else regress
until (j > 8) ∨ (j < 1);
if j > 8 then PRINT(x) else FAILURE

procedure regress;
  begin j := j − 1;
    if ¬j < 1 then
    begin removequeen;
      if i = 8 then
      begin j := j − 1;
        if ¬j<1 then
          removequeen
      end
    end
  end

procedure testsquare;
  safe := a[i] ∧ b[i+j] ∧ c[i−j]
```

```
procedure setqueen;
    a[i] := b[i+j] := c[i-j] := false

procedure removequeen;
    a[i] := b[i+j] := c[i-j] := true
```

In comparing the final programs, one sees that in Balzer's and Smith's presentations, the changes from the specifications beyond the introduction of backtracking hardly overlap at all. The major differences between Balzer's initial and final states, besides the backtracking, are the incremental updating of acceptable board positions and conversion of positions to rows, columns, and diagonals. The major difference between Smith's initial and final states, besides the backtracking, is the incorporation of the constraint prohibiting queen-captures into an easily computable method for extending partial solutions. Because the formulation of Smith's constraint is a consequence of a representational commitment in the specification only partially made in Balzer's derivation, the two derivations finish at very different places. In contrast, Wirth's presentation incorporates the major refinements of both presentations. He presents both iterative and recursive final programs, the latter of which is equivalent to Balzer's and Smith's versions. One should keep in mind, however, that both Balzer's and Smith's derivations were supported by automated systems, while Wirth's is human-oriented and much of the reasoning is informal.

7.1. Balzer

Balzer's presentation reports on work done at the University of Southern California's Information Sciences Institute on environments to support program development. This derivation was an illustration of the potential of systems based on the transformational approach to programming, as well as some practical limitations.

The derivation corresponds to a trace of a session with an interactive system in which a very knowledgeable user makes the high-level decisions and the system takes care of the details. The initial specification is a procedure generating possible solutions subject to certain constraints. First, the constraints are *unfolded* into the program, which means that checking and backtracking code is inserted wherever program actions could violate constraints. When the checks are inserted, some redundancy is introduced, which the user then removes. The user then decides to apply a transformation to unfold the backtracking, leading to an explicit loop to test all candidates. The transformation has an initially unsatisfied applicability condition that the routine containing the nondeterminism be recursive rather than iterative, so the system sets up a subgoal to satisfy this condition. A *conditioning* (previously called *jittering*) transformation is applied to produce an equivalent recursive program. Then the system checks that the transformation is complete by verifying that all consequential nondeterminism in the program has been removed. Since some nondeterminism does remain in the choice of which queen to remove when backtracking, the remaining nondeterminism is classified as incidental rather than consequential. This classification is performed to satisfy the subgoal of verification of the nondeterminism free property of the program. Some syntactic rearrangement of the program is then performed to allow the incorporate-constraint transformations to apply to the queen placement and queen-capture constraints. The effects of the conditioning include merging local variable declarations (which itself involves a conditioning subgoal), interchanging program statements, and removal of a null else clause in an if-then-else statement. Then to make the computation of the safety of a candidate position for queen placement less expensive, a finite differencing transformation is applied to compute the set of safe positions incrementally. After further simplification, a final transformation is applied to select a representation for the board position (suggested by the user) that allows efficient processing.

State	Operator	Alternative	Rationale
Specification: each queen must be placed subject to capture and occupancy constraints.		"Resulting state" specification.	[Selection] Chosen specification reduces unconventional specification constructs.
	Unfold both constraints.		[Goal] Make implicit structures explicit. [Implementation] Analysis to determine where code should be inserted.
Program with constraint enforcement code embedded.			
	Simplify.		[Goal] Remove redundancy. [Implementation] Two checks are redundant, based on variable renaming and symmetry.

(continued)

State	Operator	Alternative	Rationale
Redundant checks edited out.			
	Unfold backtracking.		[Goal] Make backtracking explicit, so it can be minimized. [Implementation] Unfold fails because program not recursive.
	Convert iteration to recursion.	Abort unfold.	[Selection] User wants unfold.
Recursive program.			
	Mark nondeterminism in queen removal as incidental.		[Goal] Needed for verification by special purpose property prover so backtracking can be unfolded. [Implementation] User knowledge of nondeterminism.
Queen removal marked as incidental; verification succeeds.			
	Assimilate queen placement constraint into generator (one queen per square).		[Goal] Reduce board positions generated.
	Merge local variables.		[Goal] Restriction must be at top level to assimilate constraint. [Implementation] Transformation to reduce nesting.
Constraint at top level.			
	Move constraint test ahead of assertions.		[Goal] Necessary for assimilation of constraint. [Implementation] Analysis of dependence of test on assertion.

(continued)

State	Operator	Alternative	Rationale
Changed test interchanged with assertion.			
	Simplify.		[Goal] Restriction is if-then-else rather than if-then, preventing assimilation. [Implementation] Null else clause not executed.
Null else clause suppressed, constraint assimilated.			
	Assimilate remaining constraint into generator (queen can't be captured).		[Goal] Further reduce board positions generated. [Implementation] Same knowledge as with other constraint.
Constraint assimilated (restriction is nested in predicate).			
	Simplify.		[Goal] Embedded conjunctions should be removed. [Implementation] Conjunction simplification.
Embedded conjunction merged with other conjunction.			
	Convert to incremental maintenance of safe positions.		[Goal] Reduce expense of determining safe positions. [Implementation] Finite differencing.
Program with incrementally calculated predicate.			

(continued)

State	Operator	Alternative	Rationale
	Simplify maintenance action.		[Goal] One of two maintenance loops is redundant.
Redundant loop removed.			
	Pick board representation.		[Goal] Minimize expense of determining safe positions. [Implementation] Representation suggested by user is efficient.
Positions represented as rows, columns, and diagonals; computation of safe positions refined to use new representation.			

The transformation system described in the presentation does provide some support for the development of the program. Yet a striking feature of this derivation is how often the user is prevented from applying transformations known to be useful by what appear to be minor syntactic mismatches to the current program state. These mismatches require the resolution of multiple levels of subgoals, and distract from the steps that are more crucial to the development, unfolding the backtracking and incorporating the constraints. They also result in the unusual step of converting an iterative program to a recursive one, which seems a step backwards in efficiency. The presentation concludes by saying that the problem is in the lack of an adequate framework for keeping track of the development. We would add that the use of a language that did not require premature commitments (to declaration of local variables and statement order, for example) might also be useful, at least in this particular derivation.

7.2. Smith

Smith's derivation demonstrates his strategy for deriving generate-and-test algorithms. The strategy is designed for implementation within a system similar to the CYPRESS system that constructed the sorting derivations, though this particular derivation was not automatically produced.

The *n*-queens problem is specified in Smith's formalism as a generator that produces solutions of a certain data type satisfying the required constraints. In general, the design strategy attempts to satisfy the specification by instantiating a scheme that recursively generates subspaces of the original space to be searched, and then finding the conditions under which solutions can be extracted when a subspace can no longer be split. For the *n*-queens problem, the specified generator produces *n*-position arrays of integers from 1 to *n* for fixed *n*. Because the height of the tree of subspaces to be searched is finite, the tree can be searched depth-first. A standard generator of solutions of this type with depth-first control is available, which leads to a backtracking structure for the algorithm. Matching this array generator to the problem provides an initial method for generating all possible assignments of queens to positions. An antecedent is derived that gives the necessary conditions for a subspace to contain solutions. This antecedent, a subspace filter, is used together with the problem input condition, the condition under which a subspace can be split, and the description of the subspaces produced by a split, to instantiate the rest of the generate-and-test scheme. The output condition of this generator is then simplified using reasoning by cases. Finally, a simple algorithm is synthesized for directly solving the problem from certain subspace descriptions: when a queen has been successfully placed in each column, then the algorithm returns its array parameter as a complete solution.

State	Operator	Alternative	Rationale
Generator specification with constraints on output array.			Given.
	Adapt standard generator with depth-first control to problem.		[Goal] Find way to enumerate output domain. [Implementation] Generator in library, function matching process, height of subspace tree is finite.
2-D array generator that places a piece at a time.			
	Derive filter on subspaces.		[Goal] Prune subspaces that can't contain solution. [Implementation] Reasoning by cases to derive consequent.

(*continued*)

State	Operator	Alternative	Rationale
Test to insure pieces placed so far are safe.			
	Simplify filter applied to initial space.		[Goal] Optimization to get guard on initial call to generator. [Implementation] Derive equivalent of initial condition.
Guard on initial call is true.			
	Synthesize generator for subspace splitting operation.		[Implementation] Generic specification for split generator; reasoning by cases; simplification for readability.
Adapted and simplified generator adds queen to partial solution.			
	Synthesize algorithms to extract solutions from subspaces.		[Implementation] Generic specification matched with true input condition.
Code to return arrays with all queens placed.			[Implementation] Generic specification is only satisfiable when all queens placed.
	Assemble algorithm.		[Goal] Produce code for input to optimization process. [Implementation] Open coding simple looping generators is an optimization.
Code with split generator open-coded.			

As with Smith's derivations of divide-and conquer algorithms, this derivation also relies on the formalization of a class of algorithms in such a way that they can be designed relatively mechanically, given the appropriate generalized deduction facilities. The idea of using subspace generators is also of interest in itself. The strategy is generally applicable since every problem can be solved by a generate-and-test algorithm.

7.3. Wirth

In his derivation of a backtracking algorithm for the 8-queens problem, Wirth explored the idea of presenting the development of a program constructed by stepwise refinement as an educational technique. The refinement follows the pattern of the other two derivations in beginning from a simple program that generates and tests all possible solutions. The strategy of preselection refers to what we now call incorporation of a constraint into a generator to generate fewer candidate solutions. In this presentation, the strategy incorporates the constraint that each column has exactly one queen into the generator. Stepwise construction of trial solutions is used to ensure that the algorithm does not try to place more queens on partial configurations with any queens that can already capture each other. An abstract backtracking program incorporating these constraints is then refined to work with concrete data representations. The representations of the chessboard and the safe squares as 8 by 8 arrays are rejected in favor of more efficient vector representations. The final steps of the presentation refine the remaining abstract operations into code and optimize the vector accesses.

State	Operator	Alternative	Rationale
Specification: Find positions so that no queens are capturable.			Given.
	Refine as generate-and-test.	Find analytical solution.	[Selection] No analytical solution known. [Implementation] Form of solution and predicate to test are known.
Program to generate all positions.			

(continued)

State	Operator	Alternative	Rationale
	Apply strategy of preselection.		[Goal] Reduce runtime by generating fewer candidates. [Implementation] Every column must contain exactly one queen.
Generate one queen per column; test by rows and diagonals.			
	Apply method of stepwise construction of trial solutions.		[Goal] Reduce runtime by pruning infeasible partial solutions. [Implementation] Only queens already placed have to be tested for mutual nonaggression.
Backtracking program that places a new queen in a column at each step.			
	Refine `trycolumn` and `regress` steps.		
`trycolumn` loop; `regress` in two steps.		Recursive `trycolumn`.	[Selection] Leads to same algorithm, choice doesn't matter.
	Choose data representation.		[Goal] Refine in direction of available programming languages.
Access procedures and board in terms of 8-position vector.		Board as 8×8 matrix.	[Selection] Simplicity of access; storage economy.
	Refine `testsquare` by introducing auxiliary variables.		[Goal] Most frequently executed instruction.

(continued)

State	Operator	Alternative	Rationale
Vectors for safe rows and diagonals.		Matrix for safe squares.	[Selection] Matrix too expensive to update when queen is removed. [Implementation] Relation of coordinate sums and differences to diagonals.
	Refine `testsquare`, `setqueen`, `removequeen`, initialization.		[Goal] Refine to code desired, ensure correctness.
All procedures as code.			
	Introduce auxiliary variable to save array access results.		[Goal] Array access occurs very often relative to index change.
Final program.			

Though not concerned with building tools to support programming, Wirth's very readable paper anticipates (in 1971!) many of the ideas that were crucial to the development of such tools. Among these ideas are: alternative refinements lead to a tree of solutions; decisions about implementation should be deferred as long as possible; it is important to use multiple design criteria (especially efficiency) in making design decisions; by reusing design effort, stepwise refinement could aid in maintaining and porting programs; and appropriate language design can support natural and clear program development. These ideas are still, in fact, the subject of much of the current research in the area.

7.4. Composite

The composite chart, Figure 7.1, incorporates the idea from all three derivations that backtracking is applicable because it is possible to incrementally generate all possible solutions in a systematic manner. Backtracking leads to a simple, but inefficient, algorithm considering all possible assignments of queens to positions on the chessboard. The algorithm's efficiency can be improved if the generator can be constrained so that it produces fewer solutions to be tested, and a useful method is to build up solutions while

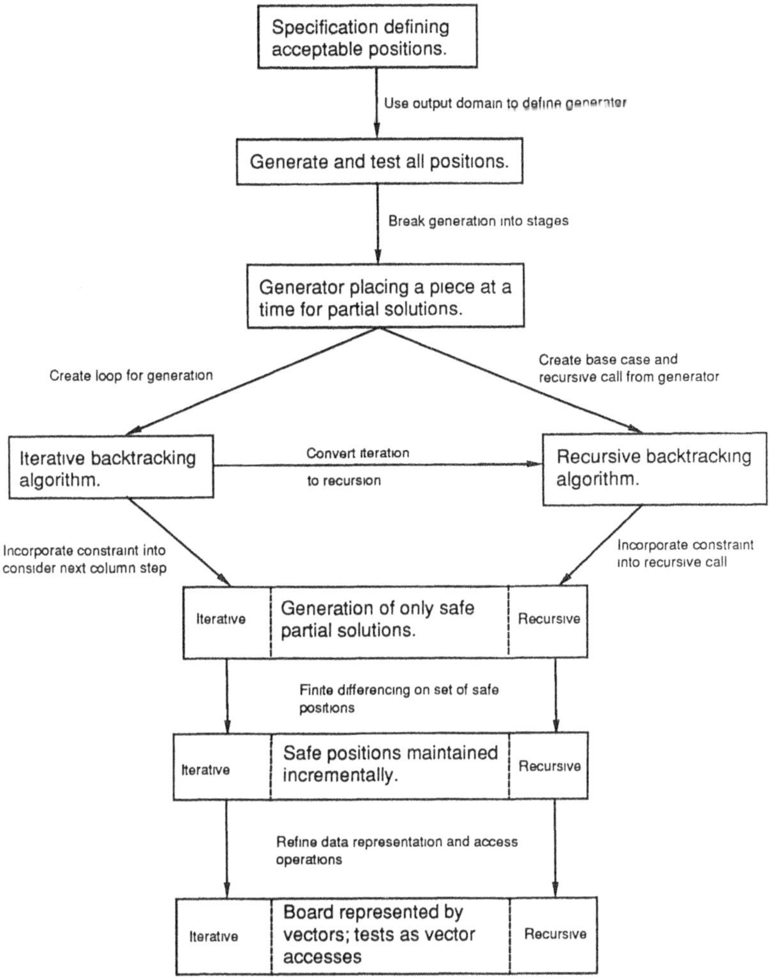

Figure 7.1. Composite for n-queens derivations.

ensuring that none of the queens placed so far are on the same row, column, or diagonal. A backtracking structure incorporating this constraint shows up in all three presentations, but in the context of different control structures. Wirth creates an iterative backtracking algorithm (though he mentions recursion as an alternative), while Smith uses the constraint to prune recursive calls to his subspace generators. Balzer creates an iterative algorithm first but then converts it to a recursive one.

A further improvement in efficiency results if the test for whether a position is acceptable is calculated incrementally. This means keeping track of the set of positions that are empty and not capturable by queens already on the board, updating the set when a square is capturable by a new queen. Balzer's derivation applies finite differencing to do this optimization, while Smith mentions that it could be done in an optimization phase. Finally, our chart shows the choice of representation of board positions as the last step, even though only Balzer's presentation made this choice last (it is part of the specification in Smith's presentation, and the result of an intermediate design choice in Wirth's). It is interesting that the presentations differed so much in when they chose the representation, indicating that the representation choices commute with the other steps in deriving this algorithm.

CHAPTER 8

Convex Hull

One of the newest subdisciplines of algorithm design is the field of computational geometry, which aims to develop efficient methods for solving geometric problems on a computer. Problems of computational geometry often arise in areas such as vision, graphics, and computer-aided design. In this chapter we will study one such problem, that of finding the convex hull of a planar set of points. The convex hull is defined as the smallest convex polygon that encloses the input point set, and it can be shown that the vertices of the convex hull form a subset of the input set. Though there are by now several efficient algorithms to solve this problem, they have been developed fairly recently, and have not been subject to the same study by the program transformation community as, say, sorting algorithms. We will examine a derivation of a divide-and-conquer convex hull algorithm from Smith [53] and the results of analyzing protocols of human algorithm designers working on the same problem from Kant and Newell [34], [36].

The two input specifications for the convex hull problem are fairly similar.

1. **Smith:** Informally the specification requires an algorithm to find a polygon (to be represented as a list of points) such that the polygon encloses the input points and its vertices are a subset of the input points. The formal specification has an implicit input condition of *true*. The output condition, and input and output domains are given below:

   Convex-hull: $b = h$ **such that** Contains: $\langle h, b\rangle \wedge$ Bag: $h \subseteq b$

   where convex-hull: Bag(POINT) $\rightarrow$ List(POINT)

2. **Kant and Newell:** Kant and Newell's designers were given an informal specification of the problem in English. The input to the algorithm is a set of points in the plane. The output can either be viewed as a convex polygon containing all the input points, whose vertices are a subset of the input, *or* as a point set comprising the vertices of a convex polygon that contains the input.

Smith's specification is equivalent to the first view of the output in Kant and Newell's specification. The latter specification contains a second view because the authors claimed that it was important that designers were able to view the type of the output not only as a convex polygon but also as a point set. The notion of being able to specify several views of the output, and allowing the algorithm designer to select the view that facilitates design most easily, was one of the novel features of the Designer system [35].

In examining the result of Smith's derivation and what Kant and Newell's designers produced, one finds that in neither case is all the processing for a divide-and conquer algorithm completely specified. Furthermore, Smith uses a conventional textual notation and Kant and Newell use a graphical one representing a dataflow language.

1. **Smith:** The output consists of a top level structure for a divide-and-conquer convex hull algorithm, and formal specifications for the composition and decomposition operators. If the input to the algorithm contains zero or one points, then a solution is computed directly by returning a list containing the input. If the input contains more than one point, it is decomposed into two bags by x-coordinate. Each of the smaller bags is recursively solved by the algorithm, and then both are composed to form the hull of the original point set.

   $$\begin{aligned} \mathsf{CH}: b \equiv\ & \textbf{if } \mathsf{Card}: b \leq 1 \\ & \quad \textbf{then } \mathsf{Listify}: b \\ & \quad \textbf{else } \mathsf{Compose} \circ (\mathsf{CH} \times \mathsf{CH}) \circ \mathsf{Decompose}: b \end{aligned}$$

   Neither the decomposition nor composition operators were synthesized in the derivation. The derivation produces a specification for the composition which has a precondition that all the x-coordinates of the points on one subhull are known to be less than or equal to the x-coordinates of the points on the other subhull, and a postcondition stating that output is a hull whose vertices are a subset of the points in the input and which contains both subhulls. The X operator applied to a bag of points returns the x-coordinates of all points in the bag. The contains relation should be understood to mean containment in the geometric sense.

   $$\begin{aligned} & \mathsf{Compose}: \langle h_1, h_2 \rangle = h_0 \\ & \quad \textbf{such that } \mathsf{X} \circ \mathsf{Bag}: h_1 \leq \mathsf{X} \circ \mathsf{Bag}: h_2 \Rightarrow \\ & \qquad\qquad \mathsf{Bag}: h_0 \subseteq \mathsf{Bag}: h_1 \cup \mathsf{Bag}: h_2 \wedge \\ & \qquad\qquad \mathsf{Contains}: \langle h_0, \mathsf{Bag}: h_1 \rangle \wedge \\ & \qquad\qquad \mathsf{Contains}: \langle h_0, \mathsf{Bag}: h_2 \rangle \end{aligned}$$

   $$\textbf{where } \mathsf{Compose}: \mathsf{List}(\text{POINT}) \times \mathsf{List}(\text{POINT}) \rightarrow \mathsf{List}(\text{POINT})$$

$$\text{Decompose}: b_0 = \langle b_1, b_2 \rangle$$
such that $\text{Card}: b_0 > 1 \Rightarrow$
$$\text{X}: b_1 \leq \text{X}: b_2 \wedge b_0 = b_1 \cup b_2 \wedge$$
$$\text{Card}: b_0 > \text{Card}: b_1 \wedge \text{Card}: b_0 > \text{Card}: b_2$$
where $\text{Decompose}: \text{Bag}(\text{POINT}) \rightarrow \text{Bag}(\text{POINT}) \times \text{Bag}(\text{POINT})$

2. **Kant and Newell:** The result of the analysis is an algorithm described as a dataflow graph, or configuration, shown in Figure 8.1. Each node in the dataflow graph represents a process component, which may be further refined into its own subconfiguration. The convex hull algorithm's divide and merge step are both refined into configurations here. Assertions on components specify processing details and arcs (links) between components define the flow of data and (implicitly) control. The algorithm has a similar structure to Smith's. The divide step is the equivalent of drawing a line through the point with the median x-coordinate. The composition step works outward from the point shared by the subhulls, repeatedly eliminating concave angles and replacing the angles by segments connecting the angle's endpoints.

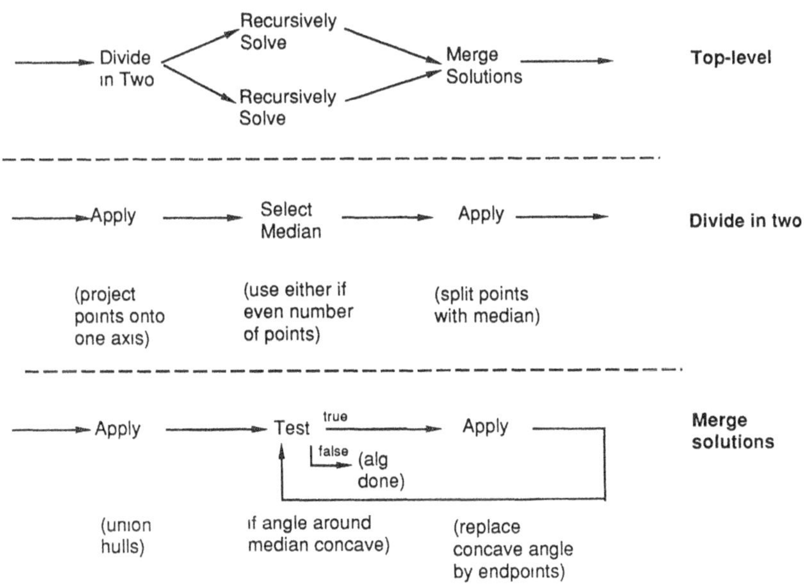

Figure 8.1. Convex-hull algorithm in Kant and Newell's dataflow notation.

8.1. Smith

Smith's derivation does not synthesize a complete convex-hull algorithm, but it illustrates that his strategy for designing divide-and-conquer algorithms works even if only the specification for decomposition or composition operators is known. The problem was treated by his semi-automatic algorithm design system.

The convex-hull derivation follows the same approach to designing divide-and-conquer algorithms as used in several of Smith's other derivations: the known specification of a decomposition operator is used to derive a specification for the composition operator. The decomposition operator is specified to split the input into two bags of points, so that the x-coordinates of all the points in one bag are less than or equal to the x-cooordinates of the others. The output condition in the composition specification is obtained by reasoning backwards from a theorem giving sufficient conditions for the existence of a correct-divide-and-conquer algorithm instantiated with the problem's output conditions. The input condition is obtained by reasoning forwards from the theorem instantiated with the output condition of the decomposition. The control predicate, to check when the solution should be computed directly, is derived from the input condition of the decomposition operator. The decomposition operator can not be applied to inputs of size less than two, so inputs with zero or one points are just returned as a list.

State	Operator	Alternative	Rationale
Specification: Map bag of points to list of points (subset of input) representing polygon enclosing input points.			Given.
	Choose decomposition specification.		
Specification for dividing points on the x-coordinate, formal equivalent of drawing vertical line through points.			
	Choose auxiliary operator.		[Implementation] Input domain produced by decomposition is same as top level.

(continued)

State	Operator	Alternative	Rationale
Auxiliary operator is recursive call to find convex hull.			
	Construct specification for composition operator.		[Goal] Satisfy correctness conditions for divide-and-conquer. [Implementation] Reasoning backward from goal to find necessary premise, which is used as output condition.
Composition takes lists of points and produces list of points representing polygon enclosing points in both inputs.			
	Choose control predicate.		[Goal] Ensure decomposition succeeds. [Implementation] Input can only be decomposed when there are two or more points.
Invoke decomposition when input is of size two or more.			
	Choose primitive (base case) operator.		[Implementation] The hull of 0 or 1 points is represented by the list containing those points.
Primitive operator is **listify**.			

Smith's strategy and system fails to design the complete convex hull algorithm, including decomposition and composition operators, for lack of the appropriate geometric knowledge. There seems to be an implicit assumption that all the knowledge must be axiomatized in some logical formalism, but constructing such an axiomatization is very difficult. The indications are that human algorithm designers succeed by representing the knowledge differently, as discussed in the next section.

8.2. Kant and Newell

In the interests of gathering data to guide the design of an automatic algorithm design system, Kant and Newell observed computer science faculty and students in the process of designing relatively complex algorithms, including several for the convex-hull problem. What we summarize here is not the performance of any one designer, but rather a composite of the ten or so people studied (though not all were successful in designing the algorithm!). We focus on the design of the divide-and-conquer algorithm to facilitate comparison with Smith's derivation.

Some of the designers independently discovered the idea of using divide-and-conquer for the basis of a convex-hull algorithm; others were explicitly instructed to use divide-and-conquer. One person had previously created a generate-and-test algorithm, and then rejected it because it was too inefficient. The designers all decided to begin with a divide step of drawing a straight line to divide the points. This divide step was used by some because it presented itself fairly quickly in a diagram, and a merge did not, while others had previous experience with algorithms in computational geometry that suggested starting with the divide step. Recursively solving both subproblems leads to the goal of finding an algorithm to merge two hulls, but only a few subjects discovered a satisfactory algorithm. Those who did decided to split the input set of points so that one point would be shared between the two sets (and consequently would be part of the solutions to both subproblems). With this divide step, they could find a relatively simple merge procedure (there are efficient algorithms to merge disjoint hulls, but they are more difficult to design). Finally, again on the basis of reasoning with sample figures, it was discovered that hulls could be directly computed for inputs with one, two, or three points.

State	Operator	Alternative	Rationale
Specification: Find polygon or points on polygon enclosing input points.			Given.
	Characterize basic idea of solution.		[Goal] Get foundation for later design.
Divide-and-conquer.		Generate-and-test.	[Selection] Generate-and-test is inefficient.

(continued)

State	Operator	Alternative	Rationale
	Find simple divide step.	Find simple merge step.	[Selection] Divide presents itself quickly in diagram.
Divide by drawing vertical line.			[Implementation] Drawing a vertical line works in dividing a sample set of points.
	Change divide to drawing through median point.		[Goal] Change divide to facilitate merge. [Implementation] Median is efficiently computable, and shared point in subhulls facilitates merge.
Divide draws through median.			
	Find hull union for merge.		[Implementation] Diagrams show that when all concave angles eliminated, convex polygon remains.
Union repeatedly eliminates concave angles from polygons starting with point.		Generating and testing possible hull union edges.	[Selection] Generating and testing edges is too expensive.
	Solve for base case.		[Implementation] Diagram shows output equal to input in these cases.
Return input for one, two, or three points.			

The discovery of the procedure for merging the subsolutions illustrates that human designers use methods which are rather different from those used by most automatic systems. For the designers studied, the shift to drawing the dividing line through one of the points came only after failing to design a merge otherwise, and drawing lines in the figure that focused attention on a common point. After this shift, the design is strongly driven by comparison of a figure displaying hulls that share a point to the desired outcome of the merge. Designers repeatedly executed possible merges on test cases, and used efficiency knowledge to analyze the complexity of their proposed designs. One person hit upon the idea of the merge because he was reminded of an algorithm for approximate solution of the Euclidean minimum cost tour problem. The algorithm was for a different problem, but it had been accompanied by a figure similar to the one the person happened to draw for the convex hull problem. This suggests that, whatever the scheme used to retrieve the relevant algorithm design knowledge, the "indexing" used for access must be at a finer grain than just pointers from complete algorithms or problems (perhaps mental imagery plays a role). Of course, the manner in which this knowledge might be acquired, represented and accessed, in automatic design systems is still an open research isssue.

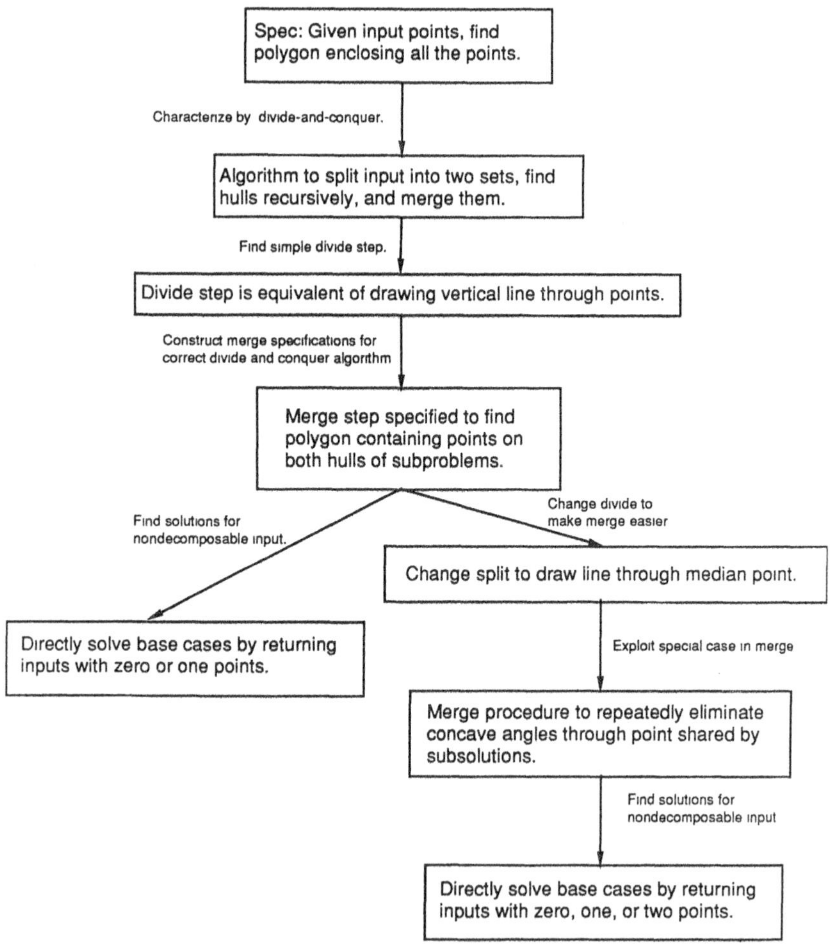

Figure 8.2. Composite for convex-hull derivations.

8.3. Composite

Much of the knowledge used in the design of this algorithm is common to both the formal derivation and the human behavior analyzed in the protocols. On the other hand, the methods used to apply the knowledge are very different. The reasoning used in Smith's derivation is very formal, while people frequently used informal reasoning: for design, test-case execution, efficiency analysis, and geometric problem-solving.

We combine the knowledge used in the presentations studied to abstract the key steps in the design, as illustrated in Figure 8.2. While a number of paradigms are applicable to this problem, use of divide-and-conquer results in very efficient algorithms. One can divide a set of points by splitting the set according to their x-coordinates, the equivalent of drawing a vertical line. This leads to another design choice (not made in Smith's presentation) about which x-coordinate is used for the split. If two roughly equal-sized subproblems are desired, one can use either the mean x-coordinate, or the median point, where the points are sorted according to x-coordinates. Either choice establishes the useful precondition that the x-coordinates of all the points in one of the inputs to the merge are less than or equal to the x-coordinates of the points in the other input. Splitting on the median point facilitates the design of the merge. The merge step finds the hull whose vertices are a subset of the union of vertices of the input subhulls and which contains (in a geometric sense) both subhulls. Whhen the two subhulls are known to share a common vertex, they may be merged by repeated elimination of the concave angles formed by the four subhull segments sharing the common endpoint. Concave angles are removed by adding a segment joining the endpoints of the angle and removing the subhull segments that formed that angle. This process continues until all concave angles have been eliminated, leaving a convex polygon. Finally, though the step to solve for base case inputs varied between the presentations, the essence is that hulls of inputs with a small number of points (three or fewer) are just all the points in the input. A complete design would then have a predicate to check for these small inputs.

CHAPTER 9

Design Space: The Final Frontier

One of the functions of a literature survey is to stimulate further research in the field. We intend the chapters on the seven algorithms to do this by presenting enough specific knowledge so that the algorithms might serve as "benchmarks" to guide the development and testing of future frameworks for design. In this final chapter we take a more general perspective, speculating on several issues in understanding the design space. The issues are divided into the general areas of goals, languages, derivation structure, implementations, and finally (an area that affected us most directly), presentation style.

9.1. Goals

The specifications in the presentations we studied consisted almost exclusively of desired input-output relationships, and correctness was defined only with respect to such relationships. For example, the sorting specifications all stated that the algorithms designed should produce an ordered permutation of the input. But functional correctness is only one design goal, and our study made it clear to us that the majority of design decisions were made based on other considerations. We examined the rationales, particularly the goal and selection knowledge, to come up with a list of such considerations. Many decisions were made because of the authors' desire to show the synthesis of a particular algorithm (where several might be possible). Other considerations (in order of decreasing frequency) were time complexity, simplicity, space complexity, and algorithm termination. Together with correctness, these categories accounted for over 90% of the goal and selection rationales. Thus, even across such widely varying presentations as the ones we examined, there is substantial regularity in the nature of the design goals.

Despite the regularity, we found that specifications, derivation systems, and techniques do not represent "nonfunctional" goals explicitly. Therefore, many factors that might influence the attainment of such goals are often not represented either. For example, in developing or choosing a sorting algorithm, one might want to answer a few questions first. These include: Will the data to be sorted fit into primary memory? What are the constraints on the space used as overhead by the sort? Should the relative order of records with identical keys be preserved, i.e., should the sort be stable? Does the input have any special characteristics that might affect the performance of the algorithm, such as being almost sorted initially, or including a large number of identical keys? Is a good worst-case performance essential?

This omission of detail cannot be attributed only to a desire to simplify the presentations. The state of the art in program specification has not reached the point where all this information can be expressed, even though it is often crucial in developing efficient programs by hand. We suggest that an interesting problem for future research would be to design specification languages capable of expressing this information. Of course, the problem remains then of how to make use of this information at design-time once it has been captured in a specification.

There are other goal-related problems not addressed in the presentations. How does one measure progress towards design goals? When goals such as efficiency, precision, or simplicity conflict, how do designers assess the tradeoffs? When would it be better to use an approximation (such as to the median in choosing the partition element in quicksort) rather than an exact algorithm? Further discussion of some of these issues, though for design in general rather than specifically for algorithm design, appears in [44].

9.2. Languages

In our desire to summarize the presentations, we have glossed over the details of particular languages used to represent specifications, partially developed programs, and final results. But it is clear to us that good language design is crucial for smoothing the derivation process. If the language does not provide appropriate abstractions, the person or system controlling the derivation will be distracted, or even overwhelmed, by details (the n-queens derivation in Section 7.1 seems to illustrate this).

Although our sources used very diverse languages, there were some abstractions that were pervasive. Nondeterminism is relied on heavily in the sorting derivations of Broy to permit one code fragment to represent a class of algorithms, and appears less crucially in the n-queens and Schorr-Waite derivations. Sets, relations, mappings, and lists are frequently used data abstractions. Abstract data types appear in only two of our sources, but their use is probably under-represented in our sample. First-order logic, functional programming, nondeterministic enumerators, and Dijkstra's guarded commands are exploited to abstract away from control flow issues.

9.3. Derivation Structure

In our composite derivation diagrams, we have represented part of the algorithm design space as a directed graph of states and operators. This, or something similar, is a commonly proposed representation [31], [61], and we have used it to elucidate the connections between apparently very different derivations. The graphs also hint at a serious difficulty for automated tools for algorithm design: the high degree of freedom inherent in the design space. At a sufficiently abstract level, there are many paths a development can take from a specification to a given implementation. Moreover, there may be many equally good implementations for one specification (Broy and Pepper's variations on the Schorr-Waite algorithm are an example of this phenomenon).

The process of making the composites has also shown that there is structure in the derivations that is not fully represented by the directed graphs. One kind of structure is that of subspaces. A difficulty in creating the composites was deciding on a level of abstraction that would be informative, and sticking to it. Often, one edge in a composite graph represents several operators in one or more of our sources. A detailed graphical representation of a particular source would expand the edge as a subgraph. An examination of alternatives at this more detailed level would produce an algorithm design subspace similar in form to the space actually shown. Another kind of structure is modularity in the design. This often leads to the commutativity of steps, and this is of course evident when two paths in the graph contain the same steps in different order. In such cases, the steps could just as well be shown occurring in parallel. For example, in the development of the Schorr-Waite algorithm, the choice of representation for the control context is independent of the choice of representation for the parameter context. But the modularity is not explicitly represented by the composite graph, and an appropriate representation is not necessarily simple to achieve. At a sufficiently high level of abstraction, the steps are independent, but a particular language and derivation methodology may strongly constrain the order in a practical sense. In Broy and Pepper's derivation, for example, their recursion removal method requires that a parameterless procedure be developed first. This requirement in turn makes it highly profitable to derive the Schorr-Waite method of representing the parameter context before removing recursion. Another example of not-so-obvious modularity in an algorithm appears in the development of quicksort: while most programmers think of the partitioning procedure as requiring a partitioning element, the introduction of the partitioning element is actually an independent optimization. Smith's quicksort derivation shows that the idea of partitioning a list by value can be developed without such an element.

Another consequence of representing designs as search through spaces is the confirmation that understanding how to control the search is still mostly an open issue. Some progress has been made in using efficiency knowledge in controlling search in the later stages of synthesis, i.e., when concrete representations must be selected for abstract types [33]. But little of this has transferred to the earlier design stages of programming, since it seems harder to instantiate and select operators (e.g., conditioning, procedure formation) in those stages. To do this automatically, we may need systems that guess, and backtrack in case of error. Knowing when to backtrack is difficult, because the set of alternatives at a decision point is rarely known to exhaust all the possibilities. Some operators may need to be applied several times before they succeed, as shown in Smith's synthesis of quicksort and Balzer's synthesis for the *n*-queens algorithms.

Possible solutions to the search control problem could lie in Kant and Newell's models of human design processes [36], including the use of symbolic and test case execution to detect problems and opportunities in design. More formal approaches might assign greater roles to proofs and deduction in constructing derivations, since proofs are now mostly used to justify, rather than to suggest, operator applications. The important exception in our survey is Smith's use of antecedent derivation in constructing specifications for subproblems.

Another issue in search control is the need to understand the effects of design choices on the achievement of design goals. For example, representation choices constrain potential efficiency gains (as in the *n*-queens derivations), and the choice of parameters in forming subprocedures affects the potential simplicity of the subprocedures (shown in the Cartesian set product derivations).

The use of design schemes can be viewed as another form of search control. The choice of such a scheme strongly constrains further design choices, and gives the derivation coherence that might be lacking without such a scheme. The abstract generate-and-test scheme is often a useful starting point in the absence of other knowledge, since all algorithms can be expressed as refinements of generate-and-test structure (pointed out in [45] and [55]). In some cases, authors have formalized the necessary conditions for the existence of an algorithm based on a particular scheme, as Smith has done for simple divide-and-conquer algorithms. Other work, not covered in this survey, explores the formalization of other schemes, such as dynamic programming [23], branch-and-bound [57], and local search [41]. In [41], Lowry used local search to derive the simplex algorithm for linear programming. His framework—choosing a design scheme, represented as an abstract parametrized theory, and reformulating the problem specification so that its constraints are incorporated into an instantiated version of the theory—offers a novel approach to structuring derivations by viewing design as problem-reformulation.

A further result of making the derivation structure explicit in our presentation summaries is that we were able to identify a number of equivalent derivational patterns. Inductively characterizing the problem domain so that larger problems may be solved by combining the solutions to smaller ones (Broy's derivations) is equivalent to choosing a divide-and-conquer scheme (Smith's derivations). Green and Barstow further showed that divide-and-conquer is sometimes reducible to the transfer paradigm. Smith, Clark and Darlington, and Green and Barstow, all make roughly equivalent decisions in designing their sorting algorithms: respectively, whether to use simple known operators for the composition or decomposition; whether the intermediate result will be similar to input or output; and whether to put the main burden of the algorithm in the consumer or producer. The processes by which Clark and Darlington, Reif and Scherlis, and Manna and Waldinger reformulated problem definitions to introduce recursive calls are similar (though the technical foundations differ). The *strategy of preselection* and *constraint incorporation* both refer to the idea of using a constraint to reduce the number of candidates to be tested in generate-and-test-algorithms (elsewhere this has been referred to as *filter promotion*, or taking advantage of *preventive constraints* [6]).

9.4. Implementations

Formalizing the design process for mechanization is a goal of many of the presentations discussed in the previous chapters. About one-third of these presentations report on derivations produced partially or completely by implemented systems. These systems are:

1. **PSI** (1974–1981) [25]: An Interlisp-based automatic program synthesis system developed at Stanford University. Of the algorithms discussed in this paper, PSI could go from specification to code for insertion sort, and from algorithm description to code for reachability (depth-first search).
2. **AFFIRM** (1978) [22]: A program verifier implemented at the University of Southern California's Information Sciences Institute (USC-ISI). AFFIRM was used in the semi-automatic verification of the asserted programs of Gerhart's Schorr–Waite derivation. Key stages of the derivation (not just the final product) were selected for the machine-assisted proofs.
3. **CYPRESS** (1982–1984) [53]: A semi-automatic algorithm design system implemented in FranzLisp at the Naval Postgraduate School. Of the examples discussed in this paper, CYPRESS goes from specifications to functional programs for insertion sort, quicksort, and Cartesian set product. On the divide-and-conquer convex hull algorithm, CYPRESS reduces the top-level specification to a specification for the composition and decomposition operators. The author of CYPRESS, Doug Smith, is now implementing a successor to CYPRESS at the Kestrel Institute,

using REFINE™, a programming environment developed by Reasoning Systems [1].

4. **DEDALUS** (1977) [42]: An interactive program synthesis implemented at Stanford in QLISP, an extension of Interlisp that includes pattern-matching and backtracking facilities. Of the examples discussed in this paper, DEDALUS was able to go from specification to code in a Lisp-like language on the Cartesian set product algorithm.
5. **TI** (1981) [4]: A Lisp-based interactive transformational system developed at USC-ISI. It took formal declarative specifications for n-queens to procedural code where the remaining details to be programmed were mainly representation selection and certain optimizations.
6. **Designer** (1982–1985) [35]: An initial attempt at a fully automatic algorithm design system, implemented at Carnegie Mellon University in a combination of Lisp and OPS5 [21], based on lessons learned from the analysis of the behavior of human algorithm designers. It was able to manipulate and execute the graphical representation used in the convex hull algorithm, but was not able to design it. A successor, Designer-Soar, is a reimplementation and extension of Designer within the Soar architecture.

This is only a brief summary of these systems; a more extensive comparison might include details of system construction and runtime behavior. Such a comparison is not possible from the published literature, which does not supply most of the necessary information. One fact that is evident from the literature on PSI, DEDALUS, and Designer, is that the sizes of their knowledge bases in number of rules (admittedly an imprecise measure without a corresponding measure of rule granularity) are not very large: about 500 for PSI and 100 each for DEDALUS and Designer.

The above list of systems includes only those discussed in the presentations we surveyed, so it omits other important work in the field, such as that of Burstall and Darlington [12], and the CIP group [9]. (For a more complete list, the reader is referred to Partsch and Steinbrüggen's survey [49].) But even taking other systems into account, there is no technology into which one can routinely feed specifications as input and obtain algorithms as output, in the same way that one can routinely compile FORTRAN programs. There has been some success in coping with some of the complexity involved in mechanizing algorithm design. There are systems that can handle different design strategies on different levels, algorithms that recurse on multiple levels, and some classes of incomplete specifications (as in Smith's derivations). But there is still a long way to go.

Another way to assess progress on the mechanization problem is to classify the derivations by the type of objects in the specifications. Of the algorithms we covered, the design of those involving sequences and sets have received the highest degree of mechanization. We conjecture that this is in part due to the close syntactic correspondence between objects and actions in the specification to the objects and actions representable in the target language. When there is a substantial dissimilarity between the specification terminology and the target language, the algorithms become very difficult to derive. Similar results have been noted before, though couched in the rhetoric of artificial intelligence rather than of program transformations: for a theory formation program to succeed, its representation of the knowledge it uses must possess a natural mapping between form and content [39].

9.5. Presentation Styles

Once one accepts that the presentations in the literature we have surveyed are in fact comparable to one another, one can begin to perform many novel comparisons to find out what the real differences are. For instance, there is a substantial variation in where different authors chose to begin and end their presentations, even for the same problems. We indicate this variety visually, with the graph in Figure 9.1. The vertical axis is year of publication, and each presentation is plotted as an interval, where the right end of the interval indicates the level of description of the input specification, and the left end, the level of the description produced by the last step in the presentation. Thus for example, Kant and Newell's protocols (published in 1984) were of designers who went from an informal specification of the convex-hull problem to a description, but not an implementation, of a divide-and-conquer algorithm. In contrast, Barstow's depth first search, or reachability, synthesis (published in 1979) went from an algorithmic description to an efficient Lisp implementation. Of course, we lack precise measures for level of abstraction, so our interval estimates are necessarily rough. Moreover, we have, in general, considered informal specifications to be "more abstract" than formal ones, but informality alone does not guarantee a high level of abstraction. For instance, the sort specification of Green and Barstow is informal, but contains a commitment to the use of the transfer paradigm; thus we have classed it as an algorithm description. Nevertheless, the graph is sufficient for supporting certain generalizations, such as the claim that the field has relatively little experience with informal specifications. (We would want a larger sample population to judge if there is any reality to the trends towards higher levels of abstractions and narrower intervals shown on the graph.)

Finally, if there are any generalizations that we can make after writing this survey, they are about what makes presentations in this field easier to assimilate. It was , of course, easiest to find the starting and finishing points of a presentation when this information was collected together in one place, rather than presented piecemeal. We found it useful when the authors listed

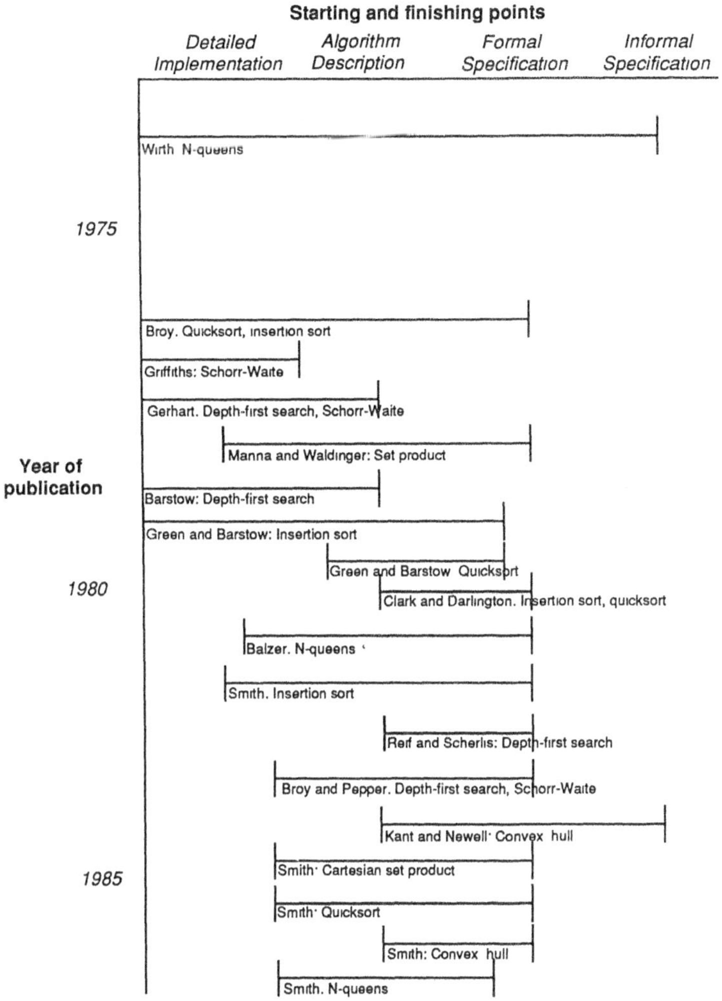

Figure 9.1. Starting and finishing points of the derivations studied.

the major design decisions in developing the algorithm, and noted alternative paths that could have been taken. Additionally, though the decision to separate design rationales into goal, selection, and implementation knowledge resulted specifically from our problem space search framework, we believe that the classification is generally helpful. Even without the classification, any clear statement of the rationale for each design decision leads to a better understanding of the presentation (as noted, and exemplified, in Wirth's paper). And last, but certainly not least, we hope that our own presentation format will encourage the use of common examples for communication between researchers in the field.

References

[1] L. Abraido-Fandiño. An overview of REFINE™ 2.0. In *Proceedings of the Second International Symposium on Knowledge Engineering*, April 1987.

[2] A. Aho, J. Hopcroft, and J. Ullman. *The Design and Analysis of Computer Algorithms*. Addison-Wesley, Boston, MA, 1974.

[3] J. Backus. Can programming be liberated from the von Neumann style? A functional style of programming and its algebra of programs. *Communications of the ACM*, 21(8):613-641, August 1978.

[4] R. Balzer. Transformational implementation: An example. *IEEE Transactions on Software Engineering*, SE-7(1):3-14, January 1981.

[5] R. Balzer, N. Goldman, and D.S. Wile. Informality in program specifications. In *Proceedings of IJCAI-77*, 1977.

[6] D.R. Barstow. The roles of knowledge and deduction in algorithm design. In A.W. Biermann, editor, *Automatic Program Construction Techniques*, chapter 10, pages 201-222, Macmillan, New York, 1984.

[7] D.R. Barstow. *Knowledge-Based Program Construction*. North-Holland, Amsterdam, 1979.

[8] J. Bates and R. Constable. Proofs as programs. *ACM Transactions on Programming Languages and Systems*, 7(1):113-136, January 1985.

[9] F. L. Bauer, H. Partsch, P. Pepper, and H. Wössner. *Notes on the Project CIP: Outline of a Transformation System*. Technical Report TUM-INFO-7729, Technische Universität München, Institut für Informatik, 1977.

[10] M. Broy. *A Case Study in Program Development: Sorting*. Technical Report TUM-INFO-7831, Technische Universität München, Institut für Informatik, November 1978.

[11] M. Broy and P. Pepper. Combining algebraic and algorithmic reasoning: An approach to the Schorr-Waite algorithm. *ACM Transactions on Programming Languages and Systems*, 4(3):362-381, 1982.

[12] R. Burstall and J. Darlington. A transformational system for developing recursive programs. *Journal of the ACM*, 24(1):44-67, 1977.

[13] K. Clark and J. Darlington. Algorithm classification through synthesis. *Computer Journal*, 23(1):61-66, 1980.

[14] R.L. Constable et al. *Implementing Mathematics with the Nuprl Proof Development System*. Prentice-Hall, Englewood Cliffs, NJ, 1986.

[15] J. Darlington. A synthesis of several sorting algorithms. *Acta Informatica*, 11(1):1-30, 1978.

[16] N. Dershowitz. The Schorr-Waite algorithm revisited. *Information Processing Letters*, 11(3):141-143, 1980.

[17] E.W. Dijkstra. A constructive approach to the problem of software correctness. *BIT*, 8:174-186, 1968.

[18] E.W. Dijkstra. *A Discipline of Programming*. Prentice-Hall, Englewood Cliffs, NJ, 1976.

[19] R.G. Dromey. *How to Solve It by Computer*. Prentice-Hall, Englewood Cliffs, NJ, 1982.

[20] M.S. Feather. A survey and classification of some program transformation approaches and techniques. In *Proceedings of the IFIP TC 2 Working Conference on Program Specification and Transformation*, Bad Tolz, West Germany, April 1986.

[21] C.L. Forgy. *OPS5 User's Manual*. Technical Report CMU-CS-81-135, Carnegie Mellon University, Computer Science Department, July 1981.

[22] S.L. Gerhart. A derivation-oriented proof of the Schorr-Waite graph marking algorithm. In F.L. Bauer and M. Broy, editors, *Lecture Notes of the International Summer School on Program Construction, Marktoberdorf 1978*, pages 472-492, Springer-Verlag, Berlin, 1979.

[23] A. Goldberg. *Transformational Derivation of Dynamic Programming Algorithms*. Technical Report KES.U.83.4, Kestrel Institute, June 1984.

[24] A.T. Goldberg. Knowledge-based programming: a survey of program design and construction techniques. *IEEE Transactions on Software Engineering*, SE-12(7), July 1986.

[25] C.C. Green, R. Gabriel, E. Kant, B. Kedzierski, B. McCune, J. Phillips, S. Tappel, and S. Westfold. Results in knowledge-based program synthesis. In *Proceedings of IJCAI-79*, pages 342-344, Tokyo, Japan, August 1979.

[26] C.C. Green and D.R. Barstow. On program synthesis knowledge. *Artificial Intelligence*, 10:241-279, 1979.

[27] D. Gries. The Schorr-Waite graph marking algorithm. In F.L. Bauer and M. Broy, editors, *Lecture Notes of the International Summer School on Program Construction, Marktoberdorf 1978*, pages 58-69, Springer-Verlag, Berlin, 1979.

[28] D. Gries. *The Science of Programming*, Springer-Verlag, Berlin, 1981.

[29] M. Griffiths. Development of the Schorr-Waite algorithm. In F.L. Bauer and M. Broy, editors, *Lecture Notes of the International Summer School on Program Construction, Marktoberdorf 1978*, pages 464-471, Springer-Verlag, Berlin, 1979.

[30] A. Hansson. *A Formal Development of Programs*. Technical Report, Dept. of Computer Science, University of Stockholm, 1980.

[31] F.A. Hussain, P. de Groote, R. Jacquart, S. Jähnichen, T.T. Nguyen, M. Sintzoff, and M. Weber. *Requirements and Feasibility Studies for a Development Language*. Technical Report RR 86-06, Unite D'Informatique Universite Catholique de Louvain, May 1986.

[32] H.B.M. Jonkers. *Deriving Algorithms by Adding and Removing Variables*. Technical Report IW 34/80, Department of Computer Science, Mathematisch Centrum, Amsterdam, April 1980.

[33] E. Kant. *Efficiency in Program Synthesis*. UMI Research Press, 1981.

[34] E. Kant. Understanding and automating algorithm design. *IEEE Transactions on Software Engineering*, SE-11(11):1361-1374, November 1985.

[35] E. Kant and A. Newell. An automatic algorithm designer: An initial implementation. In *Proceedings of AAAI-83*, American Association for Artificial Intelligence, 1983.

[36] E. Kant and A. Newell. Problem-solving techniques for the design of algorithms. *Information Processing and Management*, 20(1-2), Spring 1984.

[37] D. Knuth. *The Art of Computer Programming*. Adison-Wesley, Boston, MA, 1973.

[38] W.T. Laaser. *Synthesis of Recursive Programs*. Ph.D. thesis, Stanford University, 1979.

[39] D. Lenat and J.S. Brown. Why AM and Eurisko appear to work. *Artificial Intelligence*, 23:269-294, 1983.

[40] P.E. London and M.S. Feather. Implementing specification freedoms. *Science of Computer Programming*, 2(2):91-131, 1982.

[41] M.R. Lowry. Algorithm synthesis through problem reformulation. In *Proceedings of AAAI-87*, pages 432-436, American Association for Artificial Intelligence, 1987.

[42] Z. Manna and R. Waldinger. Synthesis: dreams → programs. *IEEE Transactions on Software Engineering*, SE-5(4):294-328, 1979.

[43] D.J. Mostow. Why are design derivations hard to replay? In *Machine Learning: A Guide to Current Research*, pages 213-217, Kluwer Academic, 1986.

[44] J. Mostow. Towards better models of the design process. *AI Magazine*, 6(1), Spring 1985.

[45] A. Newell. Heuristic programming: Ill-structured problems. In J. Aronofsky, editor, *Progress in Operations Research*, pages 360-414, Wiley, New York, 1969.

[46] A. Newell. Reasoning, problem solving and decision processes: The problem space as a fundamental category. In *Attention and Performance VIII*, Chapter 35, pages 693-718, Lawrence Erlbaum and Associates, 1980.

[47] A. Newell and H.A. Simon. *Human Problem Solving*. Prentice-Hall, Englewood Cliffs, NJ, 1972.

[48] H. Partsch. Structuring transformational developments: A case study based on Earley's recognizer. *Science of Computer Programming*, 4:17-44, 1984.

[49] H. Partsch and R. Steinbrüggen. Program transformation systems. *Computing Surveys*, 15(3), September 1983.

[50] J.H. Reif and W.L. Scherlis. *Deriving Efficient Graph Algorithms*. Technical Report CMU-CS-82-155, Carnegie Mellon University, Dept. of Computer Science, December 1982.

[51] H. Schorr and W.M. Waite. An efficient machine-independent procedure for garbage collection in various list structures. *Communications of the ACM*, 10(8):501-506, August 1967.

[52] R. Sedgewick. *Algorithms*. Addison-Wesley, Boston, MA, 1983.

[53] D.R. Smith. The design of divide-and-conquer algorithms. *Science of Computer Programming*, 5:37-58, 1985.

[54] D.R. Smith. *Top-Down Synthesis of Simple Divide and Conquer Algorithms*. Technical Report NPS52-82-011, Naval Postgraduate School, November 1982.

[55] D.R. Smith and T.T. Pressburger. *Knowledge-Based Software Development Tools*. Technical Report KES.U.87.6, Kestrel Institute, June 1987.

[56] D.R. Smith. Top-down synthesis of divide-and-conquer algorithms. *Artificial Intelligence*, 27(1):43-96, 1985. (Reprinted in *Readings in Artificial Intelligence and Software Engineering*, Editors, C. Rich and R. Waters, Morgan Kaufmann, Los Altos, CA, 1986, 35-61).

[57] D.R. Smith and T.T. Pressburger. Knowledge-based software development tools. In P. Brereton, editor, *Software Engineering Environments*, Ellis Horwood, Chichester, 1987.

[58] D.M. Steier and E. Kant. The roles of execution and analysis in algorithm design. *IEEE Transactions on Software Engineering*, SE-11(11):1375–1386, November 1985.

[59] J. Traugott. Deductive synthesis of several sorting algorithms. In *Eighth Conference on Automated Deduction*, 1986.

[60] R.C. Waters. *Automatic Analysis of the Logical Structure of Programs*. Technical Report AI-TR 492, MIT, December 1978 (Ph.D. thesis).

[61] N. Wirth. Program development by stepwise refinement. *Communications of the ACM*, 14(4):221–227, April 1971.

Index

GPSR Compliance

The European Union's (EU) General Product Safety Regulation (GPSR) is a set of rules that requires consumer products to be safe and our obligations to ensure this.

If you have any concerns about our products, you can contact us on ProductSafety@springernature.com

In case Publisher is established outside the EU, the EU authorized representative is:

Springer Nature Customer Service Center GmbH
Europaplatz 3
69115 Heidelberg, Germany

Batch number: 10712232

Printed by Printforce, the Netherlands